# Fundamentals of Materials Engineering- A Basic Guide

Authored by

**Shashanka Rajendrachari**

*Department of Metallurgical and Materials Engineering*
*Bartin University*
*Bartin-74100*
*Turkey*

&

**Orhan Uzun**

*Rectorate of Bartin University*
*Bartin-74100*
*Turkey*

# Fundamentals of Materials Engineering- A Basic Guide

Authors: Shashanka Rajendrachari and Orhan Uzun

ISBN (Online):  978-981-14-8922-8

ISBN (Print): 978-981-14-8920-4

ISBN (Paperback): 978-981-14-8921-1

need for a court order if at any point you breach any terms of this License Agreement. In no event will any delay or failure by Bentham Science Publishers in enforcing your compliance with this License Agreement constitute a waiver of any of its rights.

3. You acknowledge that you have read this License Agreement, and agree to be bound by its terms and conditions. To the extent that any other terms and conditions presented on any website of Bentham Science Publishers conflict with, or are inconsistent with, the terms and conditions set out in this License Agreement, you acknowledge that the terms and conditions set out in this License Agreement shall prevail.

**Bentham Science Publishers Pte. Ltd.**
80 Robinson Road #02-00
Singapore 068898
Singapore
Email: subscriptions@benthamscience.net

# CONTENTS

# PREFACE

In my opinion, a Materials Engineering textbook must include important topics like atomic bonding, crystal structure, imperfections, mechanical properties, polymers, composites, powder metallurgy in order to understand the basic phenomena of materials. This book covers all the above mentioned basic topics and helps students understand the structure-property relationship, characterization techniques, and applications in detail. This book can be proposed to use as an introductory course for one or two-semester in the Materials Engineering/Science program. This book is designed for all the Materials Engineering/Science students at the bachelor's, masters and PhD levels.

This book consists of 8 chapters in total related to atomic bonding, crystal structures, imperfections, mechanical properties of materials, polymers, powder metallurgy, corrosion, and composites. The main aim of the textbook is to maintain proper equilibrium between simple explanation and subject knowledge and we are successful in doing so. This book is meant for beginners who want to learn basic concepts of materials engineering. This book describes each topic in a simple and effective way. The authors of the book are more experienced in the field of materials engineering and published many research articles, books, book chapters in various reputed journals and publishers. Apart from material engineers, this book is also intended to be useful for metallurgists, chemists.

## CONSENT FOR PUBLICATION

Not applicable.

## CONFLICT OF INTEREST

The author declares no conflict of interest, financial or otherwise.

## ACKNOWLEDGEMENTS

Declared none.


**Shashanka Rajendrachari**

Department of Metallurgical and Materials Engineering
Bartin University
Bartin-74100
Turkey

&

**Orhan Uzun**

Rectorate of Bartin University
Bartin-74100
Turkey

# FOREWORD

The materials engineering is an interdisciplinary subject that has no boundary between different areas of science like chemistry, biology, physics and various branches of engineering. The materials engineering helps us to understand the structure-property relationship at the atomic level. It is possible to tune the different properties of materials just by controlling their structures both at the microscale and atomic level. This area has numerous advantages and helps to understand the materials better. The study of polymers, ceramics, composites, alloys, their applications, atomic bonding, *etc.,* come under materials engineering. It is well known that everything is a material and we are living in a materialistic world. Therefore, it is obvious that any academic curriculum in modern education must include materials science/engineering. It is also very important to provide suitable textbooks to students and researchers who are all working in materials engineering.

There are a handful number of materials engineering books available, but to understand their concept, one should know the basics of materials engineering. This book provides a greater platform to the students and researchers who choose their career in materials engineering. This book is simple to understand, properly described by a simple language and specially designed for beginners. This book comprise of a total of 8 chapters related to atomic bonding, crystal structures, imperfections, mechanical properties of materials, polymers, powder metallurgy, corrosion, and composites. The authors have described basic concepts of materials engineering in an effective and simple way so that everybody can understand these concepts easily.

If you want to learn the fundamentals of materials engineering/science then this book is the right choice. It provides sound knowledge for undergraduate, postgraduate and research students to explore the greater aspects of materials. I congratulate the authors for their excellent work and wish them good luck.

**Debasis Chaira**
Department of Metallurgical and Materials Engineering
NIT Rourkela
Rourkela
India

# CHAPTER 1

# Introduction to Materials Engineering

**Abstract:** This book discusses the study of microstructures of various materials and their classifications. It also focuses on different types of chemical bonds and the arrangement of atoms in materials. It also explains how the defects or imperfections in materials control their properties and explains the different types of strengthening mechanisms used to improve the strength of materials. This book also discusses the characterization methods used to study the mechanical properties, stress-strain curves, basics of polymers, their types, properties, and applications. Basics of powder metallurgy, mechanism of sintering, advantages and dis-advantages of powder metallurgy were explained in detail. Basic principles of corrosion, types, mechanism, and corrosion control methods, fundamentals of composite materials, their types, properties and applications are discussed in this book.

**Keywords:** Chemical bond, Composite materials, Corrosion, Imperfections, Mechanical properties, Microstructure, Polymers, Powder metallurgy, Sintering, Strengthening mechanisms, Stress-strain curves.

## 1. WHAT IS MATERIALS SCIENCE AND ENGINEERING?

Materials science and Engineering is an interdisciplinary branch of science. It usually involves the fabrication of materials, investigation of properties and their characterizations and also its applications in various fields. It also implies the relationship between the structure of materials and the various properties. The size and the structure of materials play an important role in determining the different properties.

We need correct materials to fabricate any engineering materials, alloys, structures, devices, *etc.* [1]. Knowledge of the material's properties, structure and how they react can help us to fabricate the product of our desire. Materials engineering shows us how to apply knowledge to make better things and to make things better. Fig. (**1**) shows the structure of a material [1].

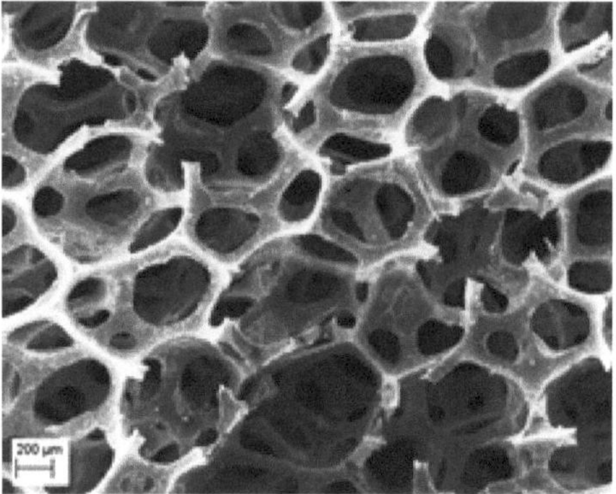

**Fig. (1).** Structure of a material [1].

This book allows students to learn more about materials science and engineering by preserving their enthusiasm throughout. The present book discusses the basic concepts of materials in a very simple manner.

## 2. WHAT EXACTLY MATERIAL ENGINEERS DO?

Material engineers will create a new type of materials by using metals, alloys, ceramics, cermets, and polymers. Usually, material engineers will do fabrication, processing, characterization and study their properties by testing materials. Material engineers are creating a huge range of products starting from micron to nano levels and they are controlling the structure of a material by using various conditions and methods. Controlling the structure and size is a big breakthrough in tailoring the properties of any materials. Materials science and engineering uplift the research in both small scale and larger scales. This will help to proceed with the academic research and industry together by transferring the technology between each other. It is fundamental to all other science and engineering disciplines [2].

If we turn back to the history of civilization, people were using stones, iron and bronze for their daily applications [3]. Unknowingly, they were using the concept of materials, like fabrication, machining, polishing, *etc.* Even today, we are using materials for various applications starting from plastic, aluminium bottles to computers, clothes, automobiles, missiles, and space technology [4, 5] as shown in Figs. (2) and (3).

## 3. CLASSIFICATION OF MATERIALS

In this chapter, we will discuss the glimpses of different types of materials, but the latter part of the book contains a detailed study. Materials can be broadly classified into mainly four types; metals, ceramics, composites, and polymers. Almost all the materials come under any one of these types and are classified mainly based upon their atomic bonding forces.

**Fig. (2).** Applications of various materials [3].

**Fig. (3).** Materials engineering from bottles to space technology [4, 5].

## 3.1. What are Metals?

Metals are typically hard, malleable, shiny, fusible and ductile solid materials. They possess good electrical, mechanical, magnetic and thermal properties. Examples of metals include iron, copper, zinc, gold, silver, *etc.* Metals can be used almost everywhere; we are in a metallic world. The application of metals starts but is not limited to domestic and industrial purposes. Noble metals like gold, silver, and platinum have a high value and are often used to make jewelry. Metals can be classified into ferrous and non-ferrous metals.

## 3.2. What are Ceramics?

The word "Ceramic" comes from the Greek word for 'pottery'. It is a non-metallic inorganic solid material composed of either metal or non-metallic compounds fabricated and shaped by different heat treatment methods at higher sintering temperatures. Generally, ceramics are hard, brittle, high-temperature and corrosion-resistant materials. These days, the "ceramic" area has been expanded to materials like glass, cement, cermet, and advanced ceramics as well. Due to their wide range of properties, ceramic materials can be used in fuel cells, ceramic filters, dental restoration, bone replacement, thermistors, dinner wares, *etc.*

## 3.3. What are Composites?

Composite materials are composed of two or more different components. They are generally made by combining two or more natural and/or artificial materials to improve the physical, chemical, electrical, thermal and mechanical properties. Composite materials are generally fabricated for a specific purpose and applications like increased hardness, electrical and thermal resistance, *etc.*

Composites are either natural or synthetic origin. Wood and cellulose are the examples of natural composites; whereas, synthetic composites are man-made. Generally, composite materials are classified into ceramic matrix, metal matrix and polymer matrix based upon the type of matrix used.

Due to their high strength, high-temperature resistance, corrosion resistance, friction resistance, and electrical resistance composite materials can be used in high-performance cookware, electrical moldings, brake system components, bearings, fire retarders, sealants, heat shield systems, *etc.*

The materials science and engineering are relatively an interdisciplinary subject connected to chemistry, biology, physics and different areas of engineering. To become a material scientist one should have a minimum bachelor's degree in materials science.

## 4. LEARNING OBJECTIVES

- To learn about the microstructural features of materials and to identify different material groups based on their microstructural features.

- To understand the different types of bonding and arrangement of atoms in crystalline structures.

- To understand the imperfections in materials and the role of these imperfections on the material's properties.

- To learn about the basic principles of different strengthening mechanisms used to improve the strength of materials.

- To learn about the mechanical properties of materials and the testing methods employed to measure these properties.

- To understand the stress-strain curves and their terminologies.

- To learn the basics of polymers, their types, properties, and applications.

- To understand the mechanism of polymer reactions, methods involved to process polymers.

- To learn about the basic principles of powder metallurgy, methods of preparation, properties, and the applications.

- To understand the mechanism of sintering and advantages of powder metallurgy.

- To learn the basic principles of corrosion, types of corrosion, mechanism of corrosion and corrosion control methods.

- To learn the fundamentals of composite materials, types, properties and applications of composite materials.

## CONCLUSION

The material engineering is an interdisciplinary subject, which deals with the study of the structure, properties and applications of different materials. Each material has different properties based upon their structures. Some of the examples for materials include polymers, ceramics, metals, alloys, composites, nanomaterials, *etc*. These materials are used almost everywhere. Literally everything is made up of materials and this world is materialistic.

## REFERENCES

[1]    Sergey Bibikov, Mikhail Prokof'Ev, and John Cuppoletti, "Composite Materials for Some Radiophysics Applications, Metal, Ceramic and Polymeric Composites for Various Uses", *IntechOpen,* no. July 20th 2011, 2011.
[http://dx.doi.org/10.5772/18074.]

[2]    "What is materials science and engineering?", https://www.sheffield.ac.uk/materials/department/what-mse

[3]     R. Friedel, "Materials That Changed History", *December 2,* 2010.  https://www.pbs.org/wgbh/nova/article/materials-changed-history/

[4]     P. Puneet, "The benefits of using the high-quality aluminum bottles!",  https://medium.com/@pathakpuneet/the-benefits-of-using-the-high-quality-aluminum-bottles- f0e5b6d0c383

[5]     "UK capital investment company initiates pioneering space technologies fund", *Room Space J.,* 2016. https://room.eu.com/news/uk-capital-        investment-company-initiates-pioneering-space-technolgies-fund

CHAPTER 2

# The Structure Of Materials

**Abstract:** Do you know all the materials are made up of atoms? And there are different types of atoms named as elements? Have you ever wondered why an atom consists of electrons, protons, and neutrons? Do you know the number of electrons in the outer shell depicts the number of atoms present? Can you tell me why metals are ductile compared to hard and brittle ceramics? Why ductile iron becomes harder when we add a small amount of carbon? Why some materials act as conductors of heat and electric current and some are insulators [1]? To understand all these concepts one should learn the atomic structure, chemical bonding, structure of amorphous and crystalline materials. The present chapter describes all these concepts in a very simple way.

**Keywords:** Amorphous materials, BCC, Covalent bonding, Crystalline materials, FCC, Ionic bonding, Lattice points, Macrostructure, Metallic bonding, Micro-structure, Nanostructure, Radius ratio, Unit cells, Van der Waals bonding.

## 1. INTRODUCTION

In this chapter, we will discuss how the microstructure, composition, and processing method of the materials control their different properties. The structure of an atom is shown in Fig. (**1**) [2].

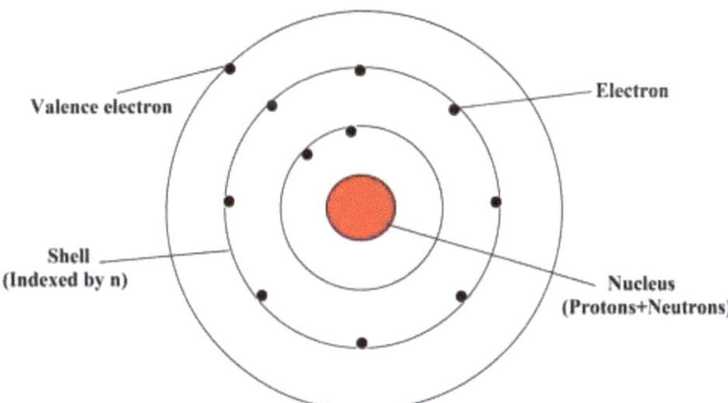


**Fig. (1).** Structure of an atom [2].

The structures of materials are classified mainly based upon their sizes; (i) atomic structure, (ii) microstructure, and (iii) macrostructure as shown in Fig. (**2**) [2 - 4].

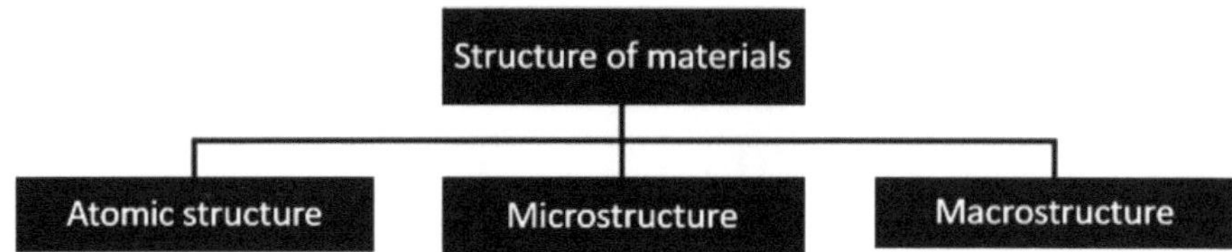


**Fig. (2).** Classification of materials.

The atomic structure consists of those features that cannot be seen through our naked eyes; like bonding between atoms and how the atoms are arranged.

As we know, the atoms are composed of electrons, protons, and neutrons, where, electrons are negatively charged particles and protons are positively charged particles, neutrons have a neutral charge. The magnitude of each charged particle in an atom is $1.6 \times 10^{-19}$ Coulombs. Both neutrons and protons have almost the same masses; whereas electrons show negligible mass. The unit of mass is an atomic mass unit (amu) $= 1.66 \times 10^{-27}$ kg and equals $1/12^{th}$ mass of a carbon atom. The Carbon nucleus has Z=6, and A=6, where 'Z' is the number of protons, and 'A' the number of neutrons. Neutrons and protons have very similar masses, roughly equal to 1 amu each [5 - 7]. A neutral atom has the same number of electrons and protons, Z.

Microstructure involves the features that can be easily seen by using an optical or electron microscope as shown in Fig. (**3**) [8]. The word "microstructure" is used to explain the appearance of the materials on the nanometer-centimeter length scale. When we describe the microstructure of the material; it is very important to consider the length scale because different length scales will give different microstructural features [8].

Macrostructure is the one that can be seen with little or no magnification or can be seen with naked eyes. Generally, the macrostructure investigations are performed on flat samples, called templates, and also on the fracture surfaces of the materials. To investigate the topography of the structure, the surface of the template is polished and carefully etched with acid or alkaline solutions. The macrostructure study reveals the discontinuity of the metal surface due to cavities, gas bubbles, porosity, *etc.*; and it also reveals the distribution of impurities, crystallites, and inclusions in different parts of the material. The macrostructure study plays an important role in determining the initial quality of the materials [9].

The chemical, physical, thermal, electrical, magnetic, and optical properties of all materials mainly depend upon their structure. Nevertheless, micro and macrostructure not only influence mechanical but also other properties. Atomic bonding affects the strength of metals by holding the atoms together by strong bonds. As we know that metals can be formable only when bonds allow atoms to move freely inside the materials. To understand this, we should discuss how the atoms are arranged and bonded in a material.

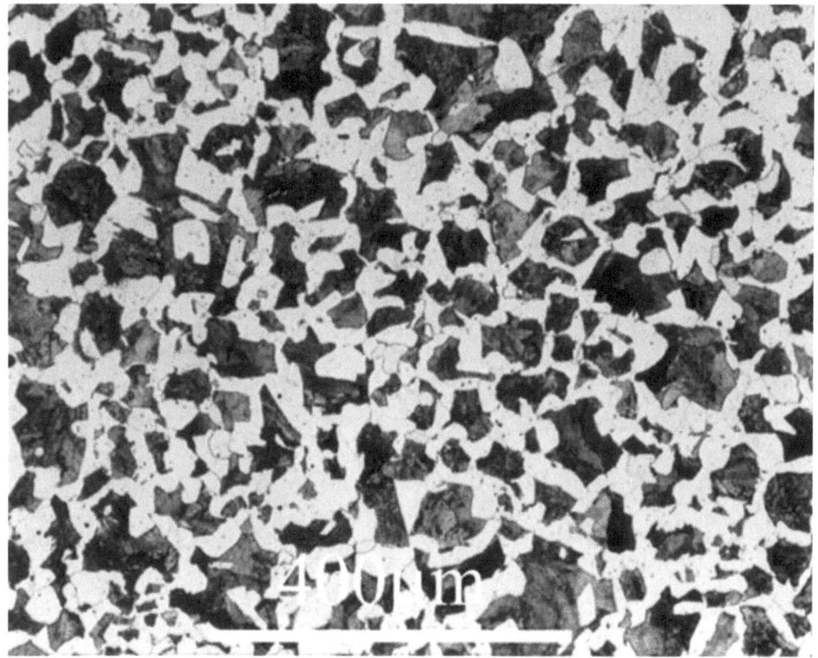

**Fig. (3).** Microstructure of steel [8].

## 2. ATOMIC BONDING

To understand the behavior of materials and wide differences in their properties, it is necessary to study their atomic level. These investigations mainly focus on the issues of how atoms are arranged and what made them constellate together. As we know, all the atoms are clustered together by different types of bonds. The inter-atomic bonds are classified into 2 types; primary bonds and secondary bonds as shown in Fig. (4). Primary bonds are very strong and they are further divided into Ionic, Covalent and Metallic bonds. Whereas secondary bonds are relatively weak and classified into Van der Waals and hydrogen bonds.

## 3. PRIMARY BONDING

### 3.1. Ionic Bonding

Each atom is made up of a specific number of protons, neutrons, and electrons; hence they are unique. Usually, the number of protons and electrons is the same for an atom. Ionic bonding is the chemical bonding formed due to the transfer of valence electrons between the two constituent atoms of the element and is non-directional [10]. An ionic bond exists between two atoms; the atom, which has an extra electron becomes a negative charge and another atom which loses an electron becomes a positive charge [11]. This results in a strong and direct Coulomb attraction.

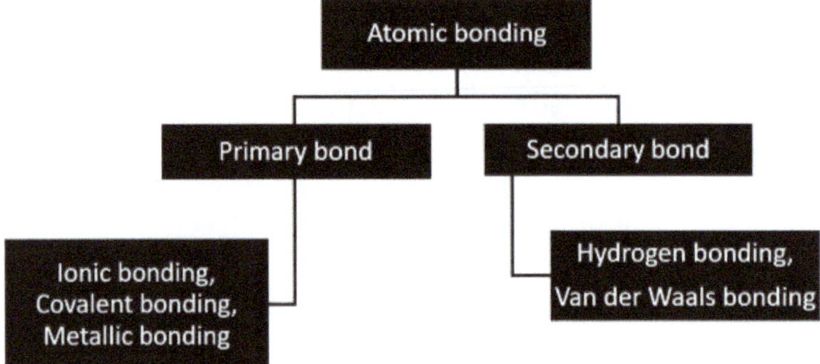


**Fig. (4).** Classification of atomic bonding.

Examples of Ionic bonds: Sodium chloride - NaCl - table salt, Calcium chloride - $CaCl_2$ – rock salt, Sodium fluoride - NaF – fluoride in toothpaste, Calcium hydroxide – $Ca(OH)_2$ – basic salt in antacid tablets.

Sodium (Na) has one valence electron in its outermost shell. It loses that one electron and becomes a $Na^{+1}$ ion. Similarly, chlorine (Cl) has 7 valence electrons in its outermost shell and it likes to gain an electron from sodium and becomes $Cl^{-1}$ ion as shown in Fig. (5) [12]. Once these atoms become $Na^{+1}$ and $Cl^{-1}$, the force of attraction between the oppositely charged ions results in the formation of an ionic bond.

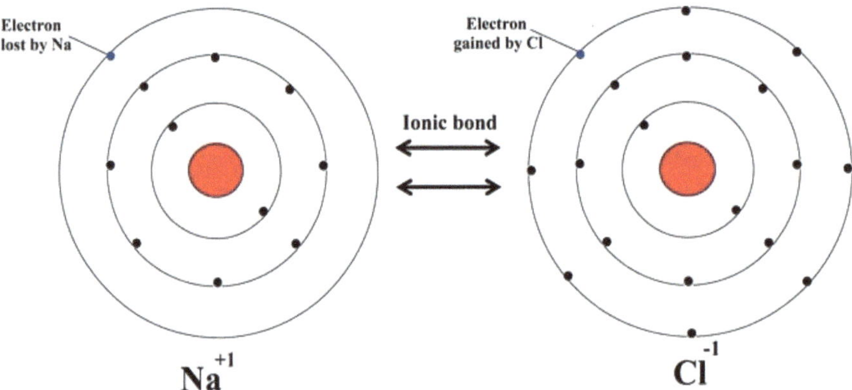


**Fig. (5).** Ionic bonding in NaCl [12].

## 3.2. Covalent Bonding

This bond is formed by the sharing of an electron pair between two atoms. This binding linkage arises from the electrostatic attraction of their nuclei for the same electrons. Covalent bonds are formed when the linked or bonded atoms have lower energy than that of widely separated. In covalent bonding, electrons are shared between the same type of atoms or different types of atoms to saturate the valency [13].

The bonded atoms prefer specific orientations relative to one another; therefore covalent bonds are directional. This directional property gives molecules a definite shape. Covalent bonding between identical atoms such as $O_2$, $H_2$ is nonpolar; hence they are electrically uniform. Whereas, the covalent bonding between unlike atoms like $CO_2$, $CH_4$ are polar; hence, one atom is slightly negatively charged and the other is slightly positively charged. This partial ionic character of covalent bonds increases with the difference in the electronegativity of the two atoms [14]. In diamond, each carbon atom is covalently bonded with 4 neighboring carbon atoms, and each neighboring atom is bonded with 4 more carbon atoms to form a rigid three-dimensional structure; therefore diamond is a very hard material.

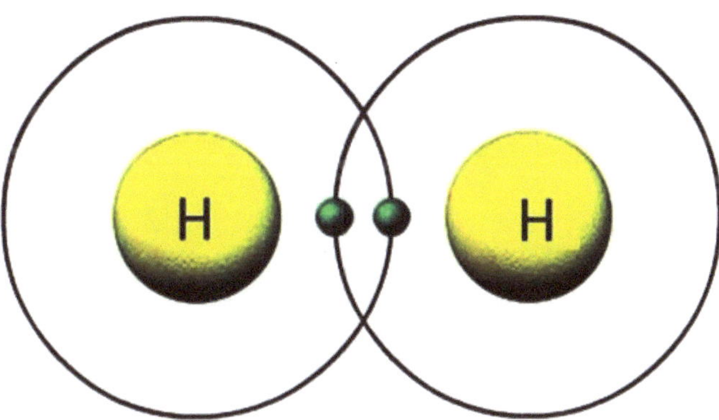


**Fig. (6).** Covalent bonding of hydrogen molecule [14].

The hydrogen molecule is formed by sharing one pair of the electron (single bond) between two hydrogen atoms as shown in Fig. (6) [14]. That is why hydrogen molecules will have a single bond.

The oxygen molecule is formed by sharing two pairs of the electron (double bond) between two oxygen atoms as shown below in Fig. (**7**). That is why the oxygen molecule will have double bonds.

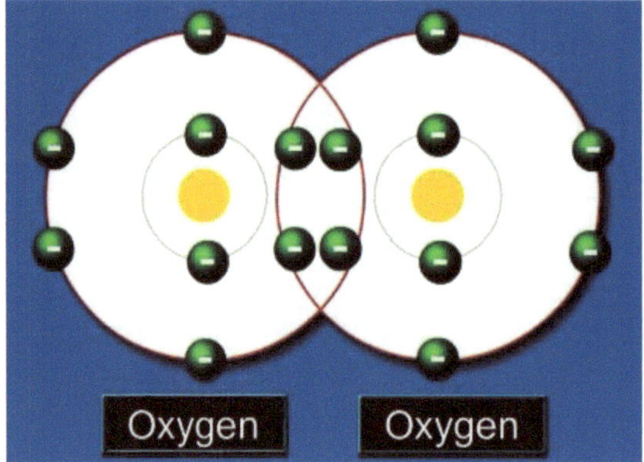


**Fig. (7).** Covalent bonding of oxygen molecule [14].

Nitrogen molecule is formed by sharing of three pairs of the electron (triple bond) between two nitrogen atoms as shown below (Fig. **8**) [14]. That is why the nitrogen molecule will have triple bonds.

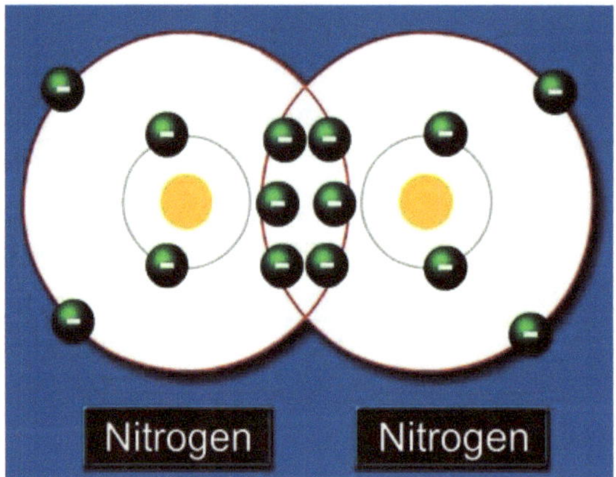


**Fig. (8).** Covalent bonding of nitrogen molecule [14].

## 3.3. Metallic Bonding

Both ionic and covalent bonds had failed to discuss the reason for electrical conductivity in metals. However, pure metals do not conduct electricity easily, and therefore they must be bonded in a different way other than an ionic and covalent bond. Metals will undergo a special type of bonding called metallic bonding [15].

In metallic bonding, the positively charged atomic nuclei share electrons freely in their electron clouds, as depicted in Fig. (9). Hence, each atom shares electrons freely with other atoms, and some of the electrons are free to move from one atom to another atom. Due to the free movement of electrons from one atom to another, the materials which are bonded by metallic bonding can conduct electric current. The force of attraction among the metal ions and the displaced electrons is very high and more amount of heat energy is required to overcome that force of attraction made the metals melt or boil. Therefore, metals will have higher boiling and melting points.

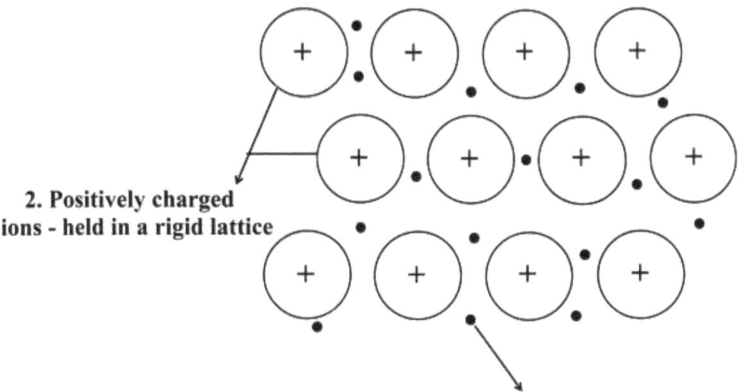


**Fig. (9).** Metallic bonding [15].

## 4. SECONDARY BONDING

Generally, secondary bonds are weak bonds and involve the attraction of partially charged atoms or molecules. When electrons concentrated on one side of an atom or molecule; then partial charges will be created to satisfy ionic or covalent bonds. This results in the formation of a polar atom or molecule which has a concentration of negative and positive charges on opposite end sides.

### 4.1. Hydrogen Bonding

Hydrogen bonding is an intermolecular force of attraction that forms a special type of dipole-dipole interaction when a hydrogen atom bonded strongly to an electronegative atom that exists in the vicinity of another electronegative atom with a lone pair of electrons. It generally occurs between covalently bonded molecules of hydrogen atoms like C-H, O-H, F-H and shares a single electron with other atom resulting in a positively charged proton that is not shielded any electrons [16]. This highly positively charged end of the molecule is capable of a strong attractive force with the negative end of an adjacent molecule. Hydrogen bonds are stronger than ordinary dipole-dipole and dispersion forces, but weaker than covalent and ionic bonds.

The properties of water are greatly influenced by these hydrogen bonds. A water molecule consists of two hydrogen atoms bonded to one oxygen atom, and they exhibit a bent structure. This is because the oxygen atom carries two pairs of unshared electrons in addition to form a bond with the hydrogen atoms. Oxygen being more electronegative and electron greedy than hydrogen, it hogs electrons and keeps them away from the hydrogen atoms. This imparts oxygen the partial negative charge at end of the water molecule; whereas, the hydrogen end shows a partial positive charge. Fig. (**10**) shows the Hydrogen bonding in water molecules [16].

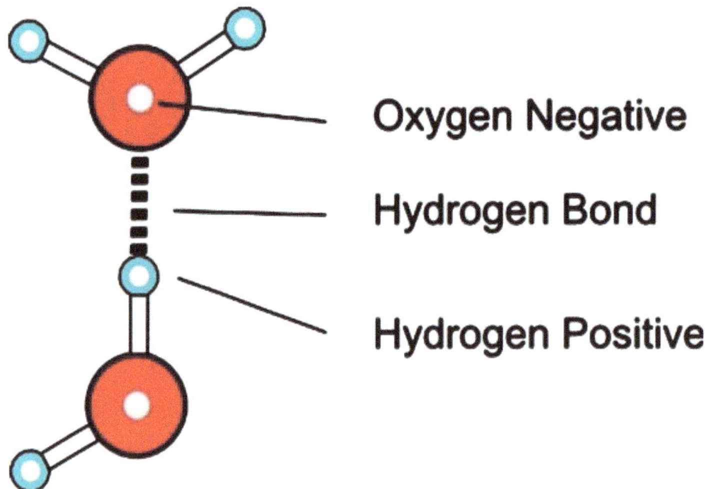


**Fig. (10).** Hydrogen bonding in water molecule [16].

## 4.2. Van Der Waals Bonding

Van der Waals bonds are secondary bonds; they are weak and result from the polarization of atoms or molecules. These types of weak electric forces attract neutral molecules to one another in gases, in liquefied and solidified gases; and also in almost all organic liquids and solids. The forces are named after the Dutch physicist Johannes Diderik van der Waals, who in 1873 first postulated these intermolecular forces in developing a theory to account for the properties of real gases. Solid materials that are held together by van der Waals forces have lower melting points and are softer than those held together by the stronger primary bonds like ionic, covalent, and metallic bonds. Fig. (**11**) represents the Van der Waals bonding [17].

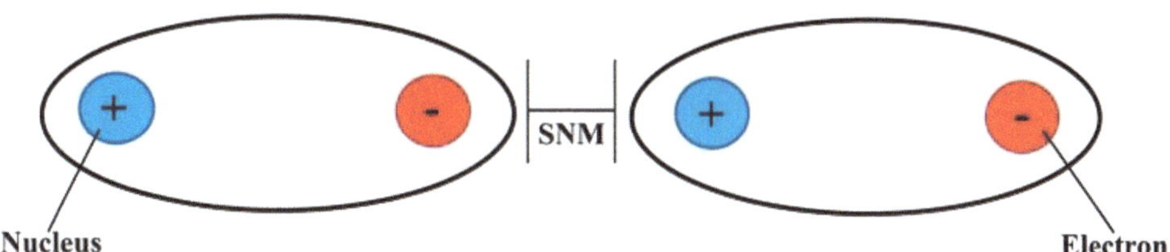


**Fig. (11).** Van der Waals bonding [17].


## 5. STRUCTURE OF METALS

The very important prospect of almost all engineering materials or metals is the way they have been constructed. In other words, most of the material's properties are strongly correlated with its structure. A good material engineer must have a sound knowledge about the relationship between structure and properties. Generally, materials can have macro, micro or nanostructures; all these structures play an important role in determining the property of a material. The microstructure is a general term used to cover a wide range of structural features, ranging from those visible to the naked eye (macrostructure) to those corresponding to the inter-atomic distances in the crystal lattice (nanostructure) [18]. To study these macro, micro, and nano-structural features, strong resolving power is needed. Optical and electron microscopes possess an adequate resolving power, using that we can very easily study the structure of materials. Fig. (**12**) depicts the size-structure diagram as follows [18, 19].

The structure in metals is of utmost importance despite its elusive manner. The majority of the metals are generally crystalline in the solid form. The metals made up of a single or same type of crystalline material are called single-crystalline metals. Whereas the large single crystals of metallic materials consist of an aggregate of different small crystals called polycrystalline materials. The crystals in the polycrystalline materials are normally called grains or crystallites.

The materials with no long-range ordered arrangement of crystals are called amorphous materials. They do not have translational symmetry and their structure is not random, but the distance between the atoms is well defined and similar to those in the crystals [20]. Fig. (**13**) shows the pictorial representation of poly and single-crystalline materials [21].

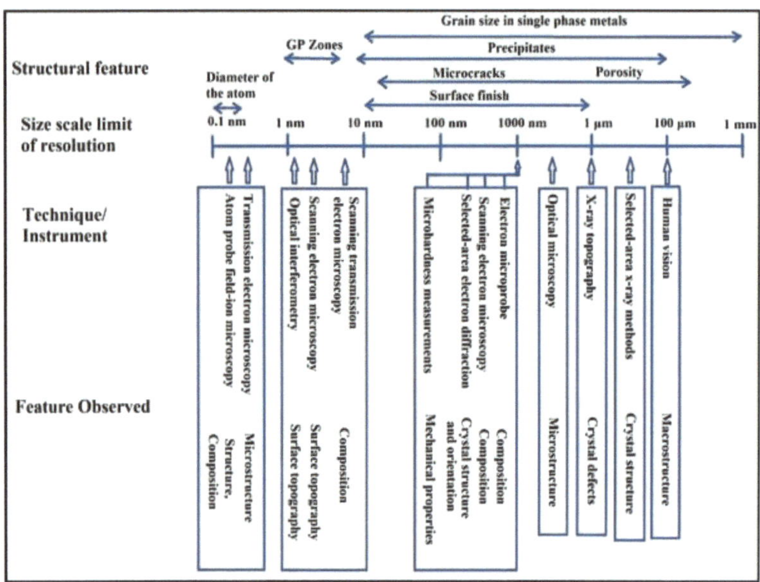

**Fig. (12).** The size scale relating structural features of metals to techniques of observation [18, 19].

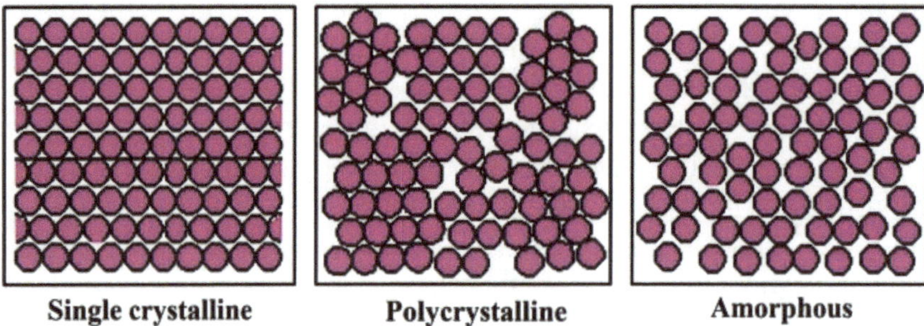


**Fig. (13).** The pictorial representation of single, polycrystalline and amorphous materials [21].

The size of the grains may vary from micron to nano range depending upon the condition of their processing. Due to their very small size, an optical microscope with higher magnification can be used to examine the topographic structural features of grains associated with metals. The structures require the range of more than 25X magnifications for their investigation will fall into the category known as microstructures. The structure of metals can be revealed by just visual examination with little or no magnification using an optical microscope are called macrostructure. Sometimes, the metallic objects after the casting process may possess very large crystals that can be easily observable to the naked eye or are sometimes easily resolved under a low power microscope; the structure in this category is termed as macrostructure. On the other hand, the materials whose grain structure or crystallite sizes are in the range of nanoscale; then they are termed as nanostructured materials (1nm=10⁻⁹m *i.e.* order of one billionth of a meter). The size scale of macro, micro, and nano-structured materials are shown in Fig. (**14**).

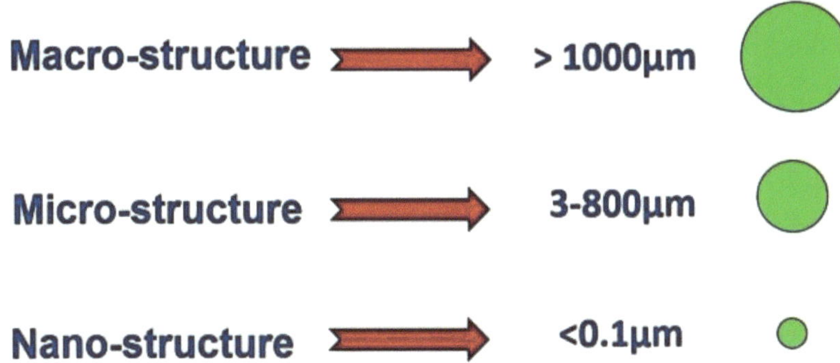


**Fig. (14).** Size scales of macro, micro, and nanostructured materials.

The basic structure inside the grains or crystallites is the atomic arrangements inside the crystals; these types of structures are called crystal structures. Among all the different forms of structure (macro, micro, and nano); the microstructure is of prime interest to the metallurgists (even nano-structure is also developing interest among the metallurgists) all over the world. This is because the metallurgical or optical microscope is generally operated at magnifications where its depth of field is extremely shallow, the metallic surface to be observed must be very flat. Meanwhile, it also must reveal the nature of the structure inside the metal accurately [18].

Most metals have crystalline structures. A crystal is defined as an ordered arrangement of a group of atoms in space. As of today, different types of crystal structures are there, some of their structures are quite complicated. However, many of the metals will crystallize in any one of the following three relatively simple structures: face-centered cubic (FCC) structure, body-centered cubic (BCC) structure, or hexagonal close-packed (HCP) structure.

## 6. LATTICE, LATTICE POINTS, MOTIF

We will discuss the lattice concept, basic concept and the unit cell concept to define the three-dimensional spatial arrangement of atoms in a crystal, and to do so, the specification of each atom's position is not required. The above concepts mainly depend upon principles of crystallography. Generally, the number of atoms present in a typical solid material is $10^{23}$ atoms per cm³. As we discussed earlier in this chapter, the structure of an atom consists of a nucleus of protons and neutrons surrounded by electrons. To easily understand the arrangement of atoms in a solid material, we will consider the atoms as hard, solid spheres; like snooker balls. Now, we will discuss the lattice and basic concept.

A lattice is a group of points called lattice points. These points are periodically arranged and each point in the lattice has identical surroundings, as shown in Fig. (15).

**Fig. (15).** Lattice points.

Generally, a lattice is assumed to be infinite in extent and this may be either having a one-, two- or three-dimensional structure. In the case of one dimension structure, there is only one lattice that is possible. Generally, points are arranged in a single line with each point are separated from each other by an equal distance. A group of one or more atoms located in a particular way concerning each other and associated with each lattice point is known as the basis or motif. The motif may contain at least one atom or many atoms of the same type or different type. We can construct a crystal structure by placing the atoms of the motif on every lattice point (*i.e.* crystal structure = lattice + motif) as shown in Fig. (16).

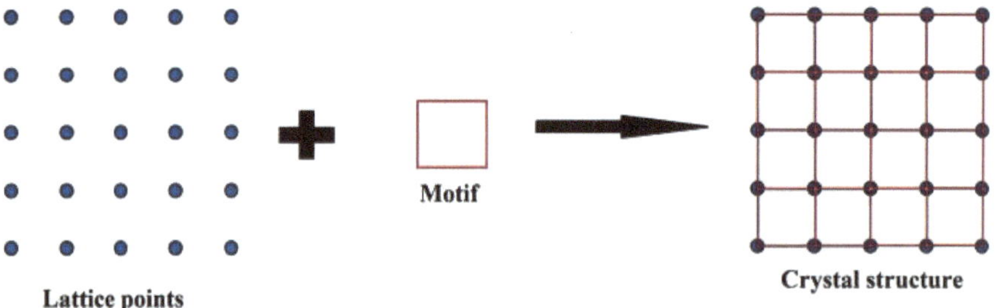


**Fig. (16).** Representation of lattice point, motif, and crystal structure.

As we discussed above, in the case of a one-dimensional structure there is only one way to arrange points such that each point has identical surroundings; an array of points separated by an equal distance. But in the case of two-dimensional structures, points can be arranged in five different ways such that each point has identical surroundings. Therefore, there are 5 two-dimensional lattices. Similarly, in the case of a three-dimensional structure, points can be arranged in fourteen unique ways. These unique three-dimensional arrangements of lattice points are known as the 'Bravais lattices'.

## 7. UNIT CELLS

The small group of atoms in a crystalline material can be arranged in a repetitive pattern. This repetitive pattern is called a unit cell. Therefore, it is more convenient to divide the crystal structure into very small repeating entities (*i.e.* unit cells) to describe any crystal structure of a solid material. The crystal structures may have square-shaped unit cells or rhombic shaped unit cells. Unit cells can be drawn within the aggregate of spheres as shown in Fig. (**17**). A unit cell is used to depict the symmetry of the crystal structure, wherein all the atom positions in the crystal may be generated by translations of the unit cell integral distances along each of its edges [4]. Thus, the unit cell is the basic structural unit or building block of the crystal structure and defines the crystal structure by its geometry and the atom positions within [4].

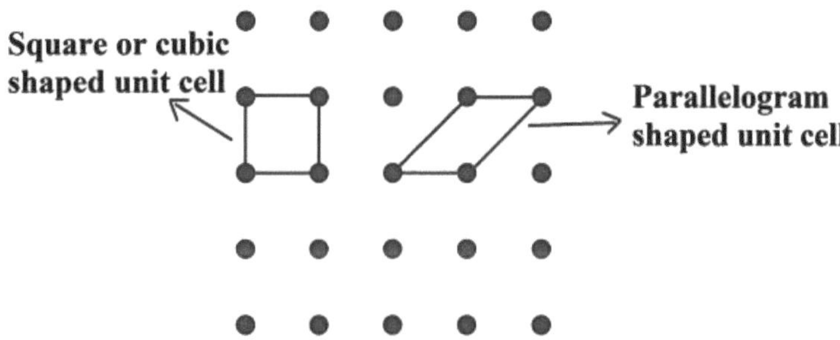


**Fig. (17).** Representation of different types of unit cells in a crystal lattice.

## 8. CRYSTAL STRUCTURES OF METALS

As we discussed earlier that metals are bonded by metallic bonding and they are non-directional. Due to this, very few restrictions will be there to the number and position of nearest-neighbor atoms. This results in a relatively large number of nearest neighbors and dense atomic packings for most of the metallic crystal structures. In the case of metals, where we use the hard-sphere model to describe the crystal structure, each sphere represents an ion core. There are mainly four types of crystal structures commonly found in most of the metals:

- Simple cubic structure
- Face centered cubic (FCC) structure
- Body-centered cubic (BCC) structure
- Hexagonal close-packed (HCP) structure

## 8.1. Simple Cubic Structure

The simple cubic structure has atoms located at each corner of the cube as shown in Fig. (**18**) [22]. Four atoms are arranged as a square at the top corner and the remaining four atoms are arranged as a square at the bottom. There are 8 eight atoms (one in each corner) for a total of ONE atom in the unit cell. This structure is called a simple cubic crystal structure. Polonium (Po) is the only metal that has a simple cubic crystal structure because the packing atoms in this way do not lead to a very high packing density.

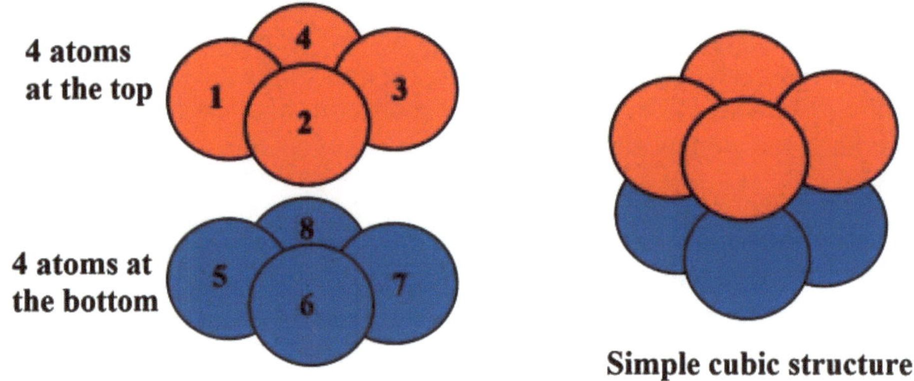


**Fig. (18).** Simple cubic crystal structure [22].

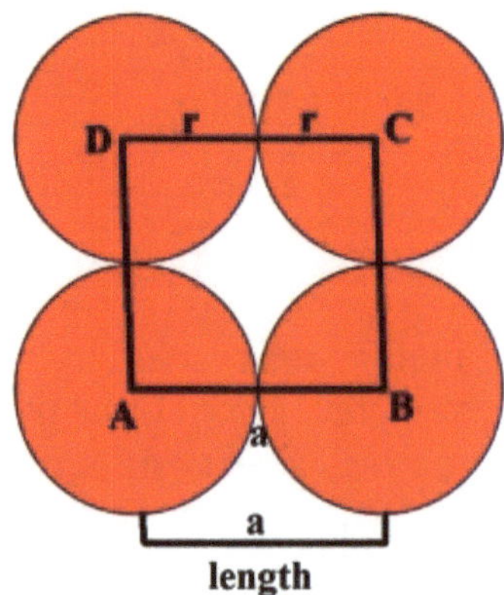


**Fig. (19).** Radius and edge length of the simple cubic structure [22].

If we count the atoms inside the cube we can see that each of these corner atoms only actually has $1/8^{th}$ of its structure in the actual cube. There are 8 corners, so the simple cubic structure contains 8×(1/8) or 1 atom per unit cell.

Consider the edge length of the cube as "a" and the atomic radius as "r". From Fig. (**19**), we can see all the sides are equal [22], so

$$AB = BC$$

$$AB = a \text{ and } BC = 2r$$

$$a = 2r$$

Then the atomic radius of simple cubic can be given as,

$$r = \frac{a}{2}$$

Atomic packing factor (APF) of simple cubic structure:

APF is the amount of volume of atoms occupied in a unit cell. APF is also called atomic packing density or atomic packing efficiency [22].

$$APF = \frac{Volume\ of\ atoms\ in\ unit\ cell}{Total\ volume\ of\ unit\ cell}$$

$$APF = \frac{Average\ number\ of\ atoms\ per\ cell \times Volume\ of\ an\ atom}{Total\ volume\ of\ unit\ cell}$$

$$APF = \frac{1 \times 4/3\pi r^3}{a^3}$$

$$APF = \frac{4/3\pi r^3}{(2r)^3}$$

$$APF = 0.52$$

## 8.2. Face Centered Cubic (FCC) Structure

Some of the metals show a face-centered cubic crystal structure. In the case of FCC, the atoms are located at each corner and on each face of the cube as shown in Fig. (**20**) [22].

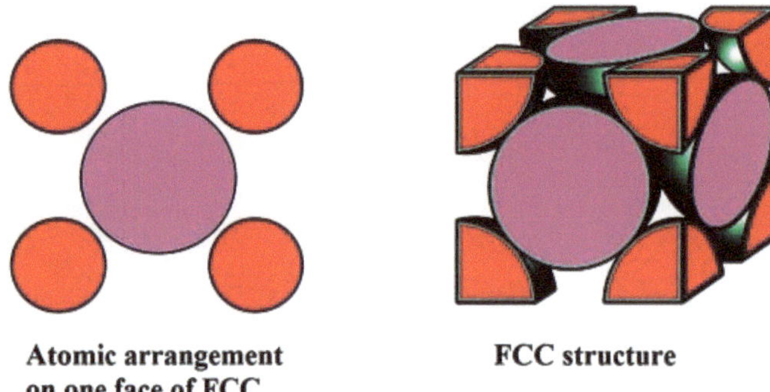



**Fig. (20).** Atomic arrangement in FCC [22].

Some of the metals such as copper, silver, aluminum, and gold have FCC structure.

Each of the corner atoms is the corner of another cube so the corner atoms are shared among eight unit cells. Therefore,

$$8 \times \frac{1}{8} = 1 \; atom$$

Additionally, each of its six face-centered atoms is shared with an adjacent atom.

$$6 \times \frac{1}{2} = 3 \; atoms$$

So the total number of atoms present in FCC,

$$8 \times \frac{1}{8} + 6 \times \frac{1}{2} = 4 \; atoms \; per \; unit \; cell$$

Consider the edge length of the cube as "a" and the atomic radius as "r". From Fig. (**21**), we can see all the sides are equal [22], so

$$(AC)^2 = (AB)^2 + (BC)^2$$

$$(4r)^2 = (a)^2 + (a)^2$$

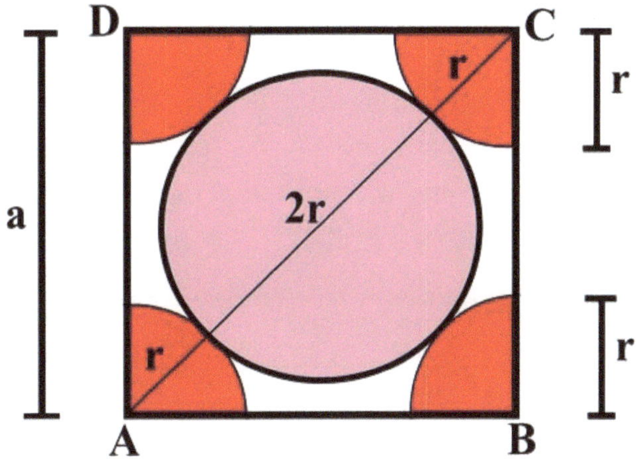

Fig. (**21**). Radius and edge length of the face-centered cubic structure [22].

Then the atomic radius of face-centered cubic can be given as,

$$r = \frac{a}{2\sqrt{2}}$$

Atomic packing factor (APF) of face-centered cubic structure:

$$APF = \frac{Average\ number\ of\ atoms\ per\ cell \times Volume\ of\ an\ atom}{Total\ volume\ of\ unit\ cell}$$

$$APF = \frac{4 \times 4/3\pi r^3}{a^3}$$

Where $r = \dfrac{a}{2\sqrt{2}}$

Substitute the value of r in the above equation, then

$$APF = \frac{4 \times \frac{4}{3}\pi(\frac{a}{2\sqrt{2}})^3}{(a)^3}$$

$$APF = 0.74$$

The metals having FCC structure at room temperature with atomic radius and lattice parameters are tabulated in Table **1**.

**Table 1. The metals having FCC structure at room temperature with atomic radius and lattice parameters [22].**

| Metal | Lattice Constant ('a' in NM) | Atomic Radius in NM |
|---|---|---|
| Aluminum | 0.405 | 0.143 |
| Nickel | 0.352 | 0.125 |
| Gold | 0.408 | 0.144 |
| Silver | 0.409 | 0.145 |
| Lead | 0.495 | 0.175 |

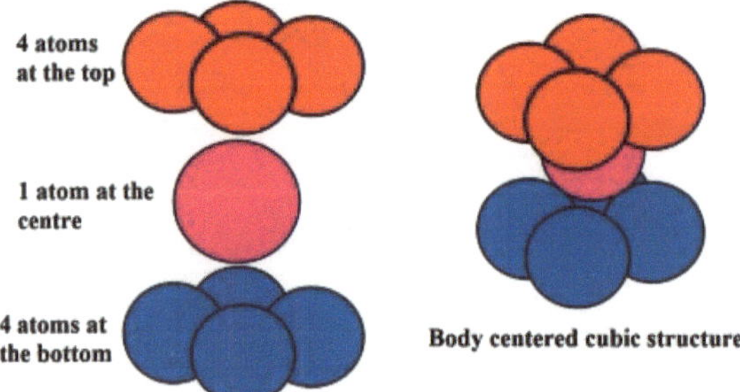


**Fig. (22).** Atomic arrangements in BCC [22].

## 8.3. Body-Centered Cubic (BCC) Structure

The body-centered cubic structure has atoms located at each corner of the cube and one at the center as shown in Fig. (**22**) [22]. Four atoms are arranged as a square at the top corner and the remaining four atoms are arranged as a square at the bottom and one atom exactly at the center.

Each of the corner atoms is the corner of another cube so the corner atoms are shared among eight unit cells. Therefore,

$$8 \times \frac{1}{8} = 1 \ atom$$

Additionally, one atom in the center of the unit cell is not sharing its atom with any adjacent atoms. Therefore we will consider that as 1 atom.

So the total number of atoms present in BCC,

$$8 \times \frac{1}{8} + 1 = 2 \ atoms \ per \ unit \ cell$$

Consider the edge length of the cube as "a" and the atomic radius as "r". From Fig. (**23**), we can see all the sides are equal [23], so

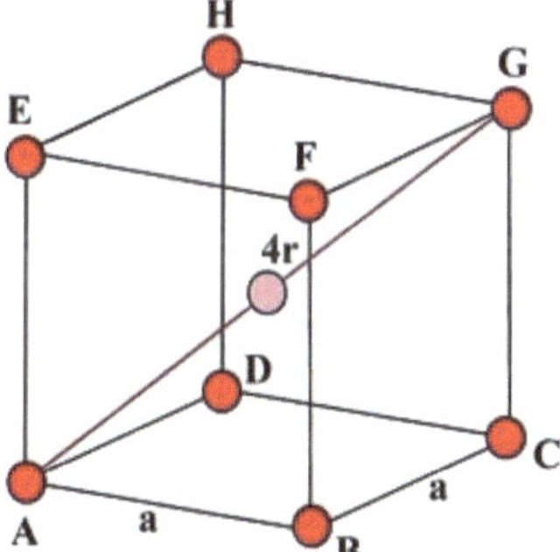


**Fig. (23).** Radius and edge length of the body-centered cubic structure [23].

$$(AG)^2 = (AC)^2 + (CG)^2$$

$$(AC)^2 = (AB)^2 + (BC)^2$$

We know that AG = 4r substitute all these values in the above equation

$$(AG)^2 = (AB)^2 + (BC)^2 + (CG)^2$$

$$(4r)^2 = (a)^2 + (a)^2 + (a)^2$$

Then the atomic radius of body-centered cubic can be given as [22, 23],

$$r = \frac{a\sqrt{3}}{4}$$

Atomic packing factor (APF) of face-centered cubic structure:

$$APF = \frac{Average\ number\ of\ atoms\ per\ cell \times Volume\ of\ an\ atom}{Total\ volume\ of\ unit\ cell}$$

$$APF = \frac{2 \times \frac{4}{3}\pi r^3}{a^3}$$

Where $r = \frac{a\sqrt{3}}{4}$

Substitute the value of r in the above equation, then

$$APF = \frac{4 \times \frac{4}{3}\pi(\frac{a\sqrt{3}}{4})^3}{(a)^3}$$

$$APF = 0.68$$

The metals having BCC structure at room temperature with atomic radius and lattice parameters are tabulated in Table **2** .

**Table 2.** The metals having BCC structure at room temperature with atomic radius and lattice parameter [22].

| Metal | Lattice Constant ('a' in NM) | Atomic Radius in NM |
|---|---|---|
| Iron | 0.287 | 0.124 |
| Sodium | 0.429 | 0.186 |
| Tungsten | 0.316 | 0.137 |
| Chromium | 0.289 | 0.125 |
| Molybdenum | 0.315 | 0.136 |

## 8.4. Hexagonal Close Packing (HCP) Crystal Structure

In the case of hexagonal close packing structure, the corner atoms are shared by a total of 6 atoms (3 atoms from above and 3 atoms from below). Similarly, the face atoms are shared by two adjacent atoms, and the atoms in the interior side are shared by only one atom as shown in Fig. **(24)** [24].

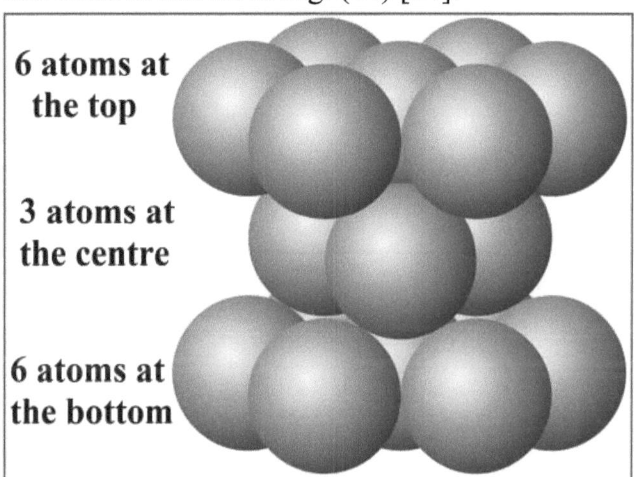


**Fig. (24).** Structure of HCP [24].

The number of atoms per unit cell in HCP is as follows;

In the case of HCP, there are a total of 12 atoms at the corner (6 atoms from the top and 6 atoms from the bottom), 2 atoms present at the center of two faces, and 3 atoms at the interior of the unit cell [24].

$$12 \times \frac{1}{6} + 2 \times \frac{1}{2} + 3 = 6 \; atom \; per \; unit \; cell$$

The structure of the HCP and the area of the unit cell ABDEFG are shown in Figs. (**25a**) and (**25b**) respectively [22]. This area is 6 times the area of the equilateral triangle ABC as represented in Fig. (**25c**) [22].

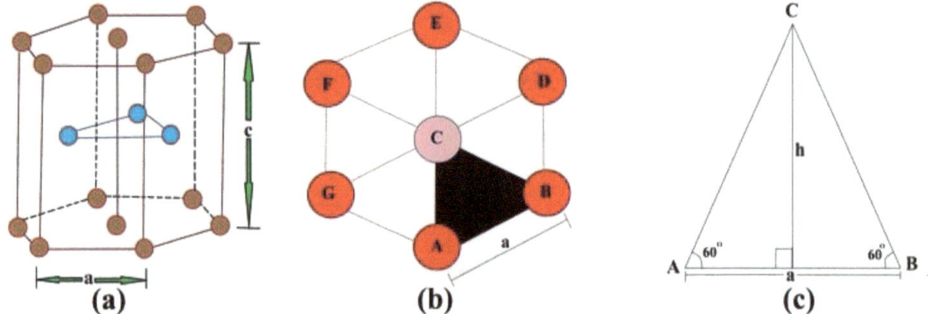

**Fig. (25).** Diagram of HCP to calculate the volume of the unit cell [22].

$$Area\ of\ \Delta\ ABC = \frac{1}{2} \times (base) \times (height)$$

But from the figure, we know that base=a, height= a sinθ

Therefore,

$$Area\ of\ \Delta\ ABC = \frac{1}{2} \times (a) \times (a\ sin\theta)$$

$$Area\ of\ \Delta\ ABC = \frac{1}{2} \times (a^2\ sin60°)$$

$$Total\ area\ ABDEFG = 6 \times \frac{1}{2} \times (a^2\ sin60°)$$

$$The\ volume\ of\ the\ unit\ cell = Area\ of\ base \times height$$

$$The\ volume\ of\ the\ unit\ cell = 3a^2\ sin60° \times c$$

From Fig. (**25b**), it is confirmed that the corner atoms A, B, D, E, F, and G are touching the center atom C from the top as well as from the bottom.

Therefore, a = 2r

Then the atomic radius of HCP can be given as,

$$r = \frac{a}{2}$$

Atomic packing factor (APF) of face-centered cubic structure:

$$APF = \frac{Average\ number\ of\ atoms\ per\ cell \times Volume\ of\ an\ atom}{Total\ volume\ of\ unit\ cell}$$

$$APF = \frac{6 \times \frac{4}{3}\pi r^3}{3a^2\ sin60° \times c}$$

Where $r = \frac{a}{2}$

Substitute the value of r in the above equation, then

$$APF = \frac{6 \times \frac{4}{3}\pi\left(\frac{a}{2}\right)^3}{3a^2\ sin60° \times c}$$

$$APF = \frac{\pi a}{3\ c\ sin60°}$$

The c/a ratio for HCP is 1.633. Substitute this value in the above equation, then;

$$APF = \frac{\pi}{3 \times 1.633 \times \ sin60°}$$

$$APF = \frac{\pi}{3 \times 1.633 \times \ 0.866}$$

$$APF = 0.74$$

The metals having HCP structure at room temperature with atomic radius, lattice parameter, and c/a ratios are tabulated in Table **3**.

**Table 3. The metals having HCP structure at room temperature with atomic radius, lattice parameter, and c/a ratios [22].**

| Metal | Lattice Constant ('a' in NM) | | Atomic Radius in NM | C/A Ratio |
|---|---|---|---|---|
| | A | C | | |
| Zinc | 0.2665 | 0.4947 | 0.133 | 1.856 |
| Cobalt | 0.2507 | 0.4069 | 0.125 | 1.623 |
| Titanium | 0.2950 | 0.4683 | 0.147 | 1.587 |
| Zirconium | 0.3231 | 0.5148 | 0.160 | 1.593 |
| Beryllium | 0.2286 | 0.3584 | 0.113 | 1.568 |

## 8.5. Radius Ratio and Co-Ordination Number

As we know, all the ionic crystals are made up of cations and anions. Anions are larger compared to cations and surround them. Both cations and anions are arranged in a space such that they touch each other and produce maximum stability [25]. The stability of the ionic crystals can be explained based on the radius ratio. Therefore, the radius ratio is the ratio of the radius of cation to the radius of an anion.

$$Radius\ ratio = \frac{Radius\ of\ cation}{Radius\ of\ anion}$$

Here, Radius of cation= r, Radius of anion = R. Thus,

$$Radius\ ratio = \frac{r}{R}$$

## 9. IMPORTANCE OF THE RADIUS RATIO RULE

This radius ratio rule is used to determine the arrangement of ions in various types of crystal structures. It also helps to determine the stability of the structure of ionic crystals. For example, the larger cations will fill the larger voids like cubic sites whereas smaller cations will fill the smaller voids such as tetrahedral sites [25]. For a particular structure, it is possible to calculate the limiting radius ratio, which is the minimum possible value for the ratio of ionic radii (r/R) for the structure to be stable. The stability of ionic compounds is shown in Fig. (26).

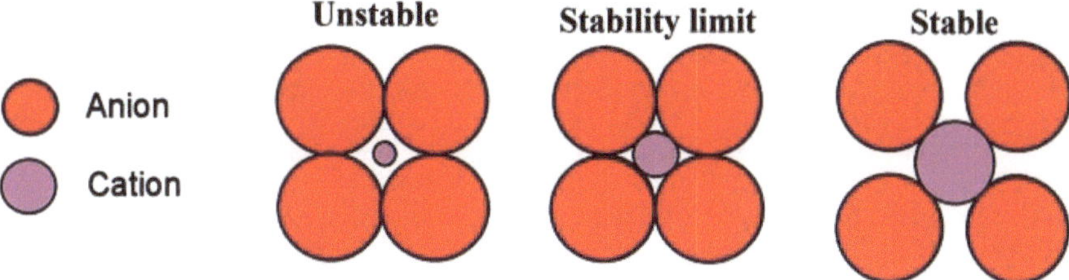


**Fig. (26).** Stability of the ionic compound to the size of cation.

It is also possible to calculate the coordination number of a compound by using the radius ratio rule.

In solid-state chemistry, the coordination number represents the number of neighbor atoms to a central atom in a crystal structure. Alfred Werner in the year 1893 coined the term co-ordination number. The coordination number for crystals and molecules is determined differently and the coordination number generally varies from as low as 2 to as high as 16. The co-ordination number calculation in a molecule or polyatomic structure is very easy and it involves just counting the number of atoms bound to it. But in the case of crystal structures, co-ordination number calculation is a little difficult and it involves counting the number of neighboring atoms; as neighboring atoms are extended in all directions in the interior of the lattice. This makes the calculation of co-ordination number a little difficult.

In BCC, the center atom is touching eight neighbor atoms (Fig. **22**). We have already discussed that all atoms in the BCC lattice are equivalent to each other. Therefore, the coordination number of the lattice is eight. The simple cubic has a coordination number of 6 and contains 1 atom per unit cell (Fig. **18**). The FCC has a coordination number of 12 and contains 4 atoms per unit cell (Fig. **20**). The radius ratio and co-ordination number of few ionic compounds are tabulated in Table **4**.

**Table 4. The radius ratio and co-ordination number of few ionic compounds [25].**

| Radius Ratio | Coordination Number | Type of Void | Example |
|---|---|---|---|
| <0.155 | 2 | Linear | |
| 0.155 – 0.225 | 3 | Triangular Planar | $B_2O_3$ |
| 0.225 – 0.414 | 4 | Tetrahedral | ZnS, CuCl |
| 0.414 – 0.732 | 6 | Octahedral | NaCl, MgO |
| 0.732 – 1.000 | 8 | Cubic | CsCl, $NH_4Br$ |
| 1 | 12 | Close packing (ccp and hcp) | metals |

## 10. DENSITY COMPUTATION

### 10.1. Volume Density ($\rho_v$)

The volume density of metal is calculated as follows,

$$Volume\ density\ (\rho_v) = \frac{Weight\ of\ atoms\ in\ the\ unit\ cell}{Volume\ of\ the\ unit\ cell}$$

$$\rho_v = \frac{Average\ number\ of\ atoms\ in\ the\ unit\ cell \times Atomic\ weight}{Avogadro's\ number\,(N) \times Volume\ of\ the\ unit\ cell}$$

$$Volume\ density\ (\rho_v) = \frac{\sum A}{NV}$$

Where, $\sum A$ = Sum of the atomic weights of atoms present in a unit cell

N = Avogadro's number ($6.023 \times 10^{23}$)

V = Volume of unit cell

The volume density of metal can be theoretically calculated by the above equation only when we consider all the atoms as hard spheres.

## 10.2. Planar Atomic Density ($\rho_p$)

The planar atomic density is the fraction of the total crystallographic plane area that is occupied by atoms. Therefore, planar atomic density is given as,

$$\rho_p = \frac{Number\ of\ atoms\ occupying\ crystallographic\ plane\ area}{Selected\ area}$$

## 11. CRYSTAL SYSTEMS

The crystal structure of solid materials generally described by using a geometric concept called space lattice. Space Lattice is a three-dimensional representation of atoms and molecules in space arranged in a specific order/pattern. This is also called the point lattice.

**Crystal:** It is a regular and repeating pattern of constituent atoms.

**Crystal lattice:** Regular 3D arrangement of points in space (Crystal lattice is not found in amorphous solids).

**Point:** Represent constituent particles (atoms, molecules, or group of atoms).

**Lattice point:** Each point in a lattice is called a lattice point or lattice site (lattice points are joined to give geometry).

## 12. CHARACTERISTICS OF THE UNIT CELL

A unit cell possesses the following characteristics [26]:

- A unit cell has three sides a, b, and c, and three angles $\alpha$, $\beta$, and $\gamma$ between the respective sides.

- The sides a, b and c may or may not be perpendicular to each other.

- The angle between edge 'a' and 'b' is $\alpha$, 'b' and 'c' is $\beta$ and that of between 'a' and 'c' is $\gamma$ as shown in Fig. (**27**) [26].

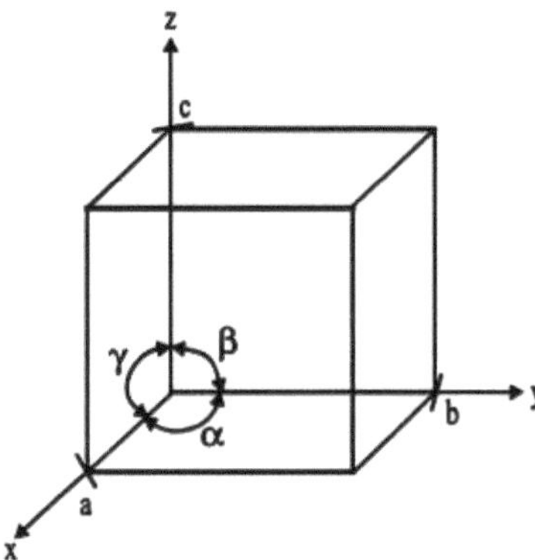

**Fig. (27).** A unit cell [26].

These side lengths (a, b, c) and angles between them ($\alpha$, $\beta$, $\gamma$) are called lattice constants or lattice parameters of the unit cell [27]. By varying the above lattice constants; we can get 7 crystal systems and they are tabulated in Table **5** and the same is represented in Fig. (**28**) [27].

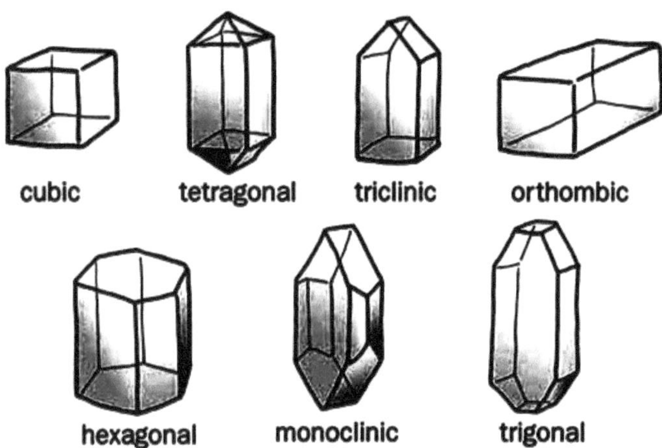

**Fig. (28).** Crystal systems [27].

A. Bravais in the year 1848 showed that; if we combine 4 unit cells (simple cubic, FCC, BCC, and edge centered cubic lattice) and 7 crystal systems; then 14 different three-dimensional space lattice are possible and called them Bravais lattices. All the 14 Bravais lattices are represented in Fig. (**29**) [28].

**Table 5. Crystal systems [27].**

| System | Side Lengths | Side Angles |
|---|---|---|
| Cubic | a=b=c | $\alpha=\beta=\gamma=90°$ |
| Tetragonal | a=b≠c | $\alpha=\beta=\gamma=90°$ |
| Orthorhombic | a≠b≠c | $\alpha=\beta=\gamma=90°$ |
| Rhombohedral | a=b=c | $\alpha=\beta=\gamma\neq90°$ |
| Hexagonal | a=b≠c | $\alpha=\beta=90°, \gamma=120°$ |
| Monoclinic | a≠b≠c | $\alpha=\gamma=90°, \beta \neq 120°$ |
| Triclinic | a≠b≠c | $\alpha \neq\beta \neq\gamma \neq 90°$ |

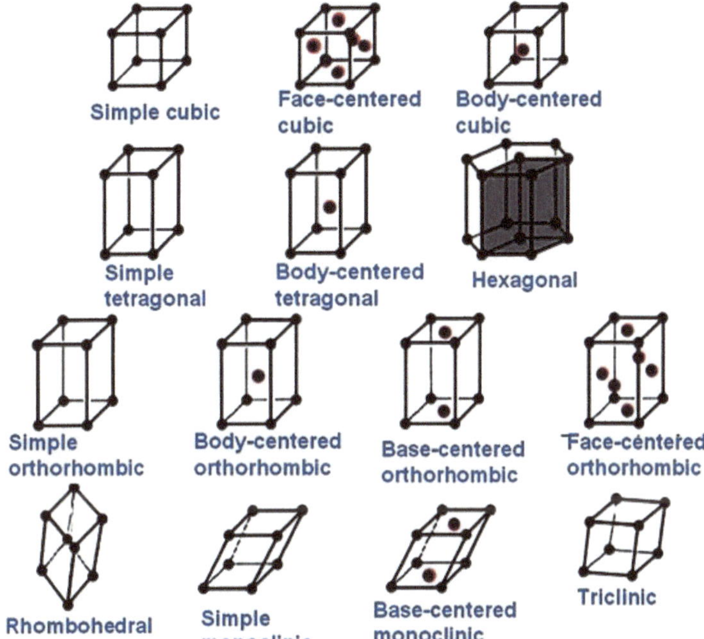


**Fig. (29).** Bravais lattices [28].

## CONCLUSION

This chapter described the different types of atomic bonding in detail and also explained how bonding affects the structure of materials. Ionic, covalent, metallic, hydrogen, and Van der Waals bonding were discussed successfully. We put extra stress on understanding the crystal structure of metals and their classifications into FCC, BCC, and HCP, respectively. We also discussed the radius ratio rule, co-ordination number, and density computation. This chapter also discussed different crystal systems and characteristics of the unit cell.

## QUESTIONS

1. Discuss the classification of the structure of materials.
2. What is meant by microstructure?
3. How the structure of materials influences their properties?
4. Discuss the classification of atomic bonding.
5. Differentiate between primary and secondary bonding?
6. What is meant by ionic, covalent, metallic, hydrogen, and Van der Waal's bonding?
7. Define the terms macrostructure, microstructure, and nanostructure.
8. Differentiate between crystalline and amorphous materials
9. Write a note on the following terms;
   a. Lattice
   b. Lattice points
   c. Unit cell
   d. Motif
10. Mention the different types of crystal structures commonly found in metals.
11. Calculate the atomic radius, atomic packing fraction for simple cubic, BCC, and FCC structures.
12. What are the radius ratio rule and co-ordination number? How they are interrelated?
13. Explain the importance of the radius ratio rule.
14. What is volume density and planar atomic density?
15. What is the characteristic of a unit cell?
16. Discuss the various crystal systems with one example each.
17. What is meant by Bravais lattices?

## NUMERICALS

1. The ionic radius of an anion is 0.234nm and that of the cation is 0.157nm. Calculate the radius ratio and predict the co-ordination number and type of void.
2. Calculate the radius ratio, ionic radius of anion, and cation for hexagonal close packing structure.
3. Calculate the volume density of FCC iron. Given that; lattice parameter of iron is 0.351nm, the atomic weight is 55.84 g/mole and Avogadro's number is $6.02\times 10^{23}$ atoms/mole.
4. Calculate the number of atoms per unit cell for simple cubic, FCC, BCC, and HCP structures.
5. Calculate the atomic radius of aluminum having a lattice parameter of 0.45nm.

# REFERENCES

[1]    https://slideplayer.com/slide/7528379/

[2]    "Structure of Materials", NTD Resource center. https://www.nde-ed.org/EducationResources/CommunityCollege/Materials/Structure/introduction.htm

[3]    S.V. Kailas. "National Programme on Technology Enhanced Learning". NPTEL, 2011.

[4]    W.D. Callister Jr, *Materials Science and Engineering – An introduction.* 6th ed. John Wiley & Sons, Inc. NJ, 2004.

[5]    B.D. Cullity, and S.R. Stock, *Elements of X-Ray Diffraction.* 3rd ed. Prentice-Hall: Upper Saddle River, NJ, 2001.

[6]    L.H. Van Vlack, *Elements of Materials Science and Engineering.* 6th ed. Addison Wesley Longman, Inc.: New York, 1998.

[7]    *Practical 17, Materials and Minerals Science.* Course C: Microstructure, CP1.

[8]    H.K.D.H. Bhadeshia, "Interpretation of the Microstructure of Steels", *Microstructure of Steels, University of Cambridge,* 2015.   http://cml.postech.ac.kr/2008/Steel_ Microstructure/SM2.html

[9]    McGraw-Hill, S. P. Parker., *McGraw-Hill Dictionary of Scientific & Technical Terms,* The McGraw-Hill Companies, Inc.: New York, 2003.  https://encyclopedia2. thefreedictionary.com/Macrostructure

[10]   S. Friedl, *Ionic compounds.*  https://study.com/academy/lesson/what-are-ionic-reactions-defini-ion-examples.html

[11]   L. Urbano, "An Introduction to Ionic Bonding", *Montessori Muddle, Retrieved June 30th,,* 2013. http://MontessoriMuddle.org/

[12]   S.A. Nelson, *Crystal chemistry, Mineralogy, Tulane University, EENS 1110,* 2015. https://www. tulane.edu/~sanelson/eens1110/minerals.pdf

[13]   *Covalent bond, The Editors of Encyclopaedia Britannica, Encyclopaedia Britannica.*

[14]   *Satyanarayana, Subject Matter Expert, covalent bond, Next. Gurukul.* https://www.nextgurukul.in/nganswers/ask-question/answer/Covalent-bond-/Carbon-and-its-Compounds/76202.htm

[15]   https://byjus.com/chemistry/metallic-bonds/

[16]   History of the Universe, http://www.historyoftheuniverse.com/index.php?p=hydrbond.htm

[17]   https://revisionscience.com/a2-level-levelrevision/chemistry-level-revision/bonding-and-structu-e/van-der-waals-forces

[18]   A. Paul, T. Laurila, V. Vuorinen, and S.V. Divinski, *Thermodynamics, Diffusion and the Kirkendall Effect in Solids.* Springer International Publishing: Switzerland, 2014. [http://dx.doi.org/10.1007/978-3-319-07461-0]

[19]   A.S.M. Handbook, *Metallography and Microstructures.* vol. 9. ASM International Materials Park: Ohio, USA, 2004.

[20]   DoITPoMS, *Atomic-Scale Structure of Materials.* University of Cambridge, 2004.

[21]   K.S. Reddy, "Metallurgy & Material Science-MMS-UNIT-1",  https://www.slideshare.net/avutu_ kunduru/ksrinivasulureddysnistmetallurgy-material-sciencemmsunit1

[22]   V.D. Kodgire, and S.V. Kodgire, *Material Science and Metallurgy for Engineers.* 26th ed. Everest Publishing House: India, 2010.

[23]   *Physics in nutshell, Solid state physics, Body-centered cubic.* https://www.physics-in-a-nutshell.com/article/12/body-centered-cubic-bcc

[24]   E. Amayuelas, A.F. Marijuan, G. Barandika, B. Bazán, M-K. Urtiaga, and M.I. Arriortua, "Mother

structures related to the hexagonal and cubic close packing in Cu24 clusters: solvent influenced derivatives", *Cryst Eng Comm,* vol. 17, pp. 3297-3304, 2015.
[http://dx.doi.org/10.1039/C5CE00251F]

[25]  *Toppr, Chemistry, Solid-state, Radius ratio rule.* https://www.toppr.com/guides/chemistry/the-solid-state/radius-ratio-rule/

[26]  T. Globe, *Tutors Globe, UNIT CELL.* http://www.tutorsglobe.com/homework-help/physical-chemistry/unit-cell-72605.aspx

[27]  "Science7Rosha, Crystal structures, What are Crystalline Structures and how do they form?", http://roshascience7.weebly.com/crystals-structures.html

[28]  Y. Liao, "Practical Electron Microscopy and Database, An Online Book", GlobalSino, 2006. https://www.globalsino. com/EM/page4546.html

**CHAPTER 3**

---

# Imperfections or Defects in Crystals

**Abstract:** Properties of most of the materials are tremendously influenced by some imperfections or defects. These imperfections greatly alter the mechanical, electrical, surface, chemical, and thermal properties of materials. Therefore, one should have complete knowledge of defects or imperfections. The mechanical properties of pure metals greatly change when they are alloyed; for example, steel (a combination of iron and 0.1% carbon) is much harder and stronger than pure iron. Bronze (20% tin and 80% copper by mass) produces a much harder and sharper weapon than pure tin or pure copper. Jewelry gold (pure gold and 5% copper by mass) is a much harder, lustrous, and durable gold than pure gold.

**Keywords:** Dispersion strengthening, Edge and screw dislocations, Frenkel defect, Grain boundary defect, Grain boundary strengthening, Line defects, Point defects, Plane defects, Precipitation hardening, Schottky defect, Solid solution hardening, Strain hardening, Twin boundary defect, Volume defects.

## 1. INTRODUCTION

Theoretically, an idea of a perfect crystal may be possible, but experimentally it is very difficult to prepare a crystal with zero defects. If materials are perfect crystals then their properties would be explained by their composition and crystal structure [1]. Due to the imperfections in the crystals; scientists can tailor the properties of materials into diverse combinations that modern engineering devices require. This chapter will describe the different types of defects present in crystalline solids and also explains how a defect can alter the material's property.

Any deviation from the periodic arrangement of atoms in an ideally perfect crystal is said to have an imperfection or a defect. Fig. (**1**) depicts the defects in solid materials. The type of imperfections, number, and extent of imperfections can affect the material's property from a lower to a greater extent. The conductivity of the semiconductors is mainly due to the defects caused by added chemical impurities. Atomic diffusion, color, and the luminescence of the various crystals mainly depend on the impurities or defects [2].

**Shashanka Rajendrachari & Orhan Uzun**

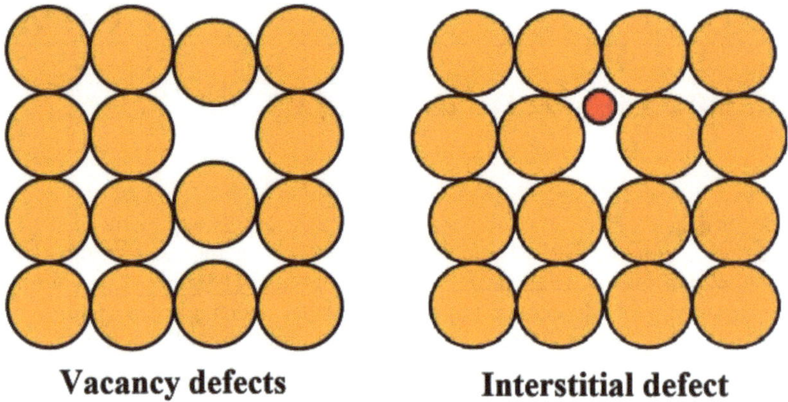



**Fig. (1).** Defects in solids.

## 2. TYPES OF DEFECTS

Imperfections or defects can be broadly classified as follows and represented in Fig. (**2**).

1. Point defects (also called Zero dimensional defects)
2. Line defects (also called one-dimensional defects)
3. Plane or surface defects (also called Two-dimensional defects)
4. Volume defects (also called three-dimensional defects)

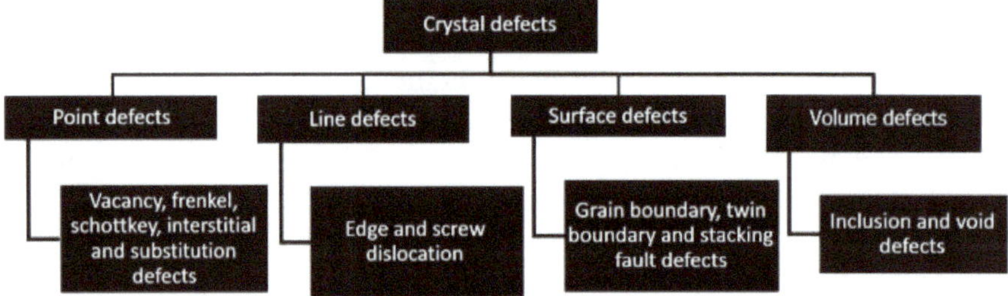



**Fig. (2).** Classification of imperfections or defects.

## 3. POINT DEFECTS

As we discussed, all the atoms have vibrational energy above their absolute zero temperature. This energy is enough to break the bonds between two atoms and then atoms will become free from their lattices. This results in a defect called point defects.

## 3.1. Vacancy Defect

The common of all the point defects is a vacancy defect. As we know, it is not possible to create a crystal without a defect. When a crystal has vacancies in its lattice, sites are called to have a vacancy defect. This defect decreases the density of the substance and it is formed when a substance is heated.

The existence of vacancies can be well explained by using thermodynamic principles. As the number of vacancies in the crystal increases; entropy will also increase. The equilibrium number of vacancies at any temperature is given as follows;

$$n_v = n_t e^{\left(\frac{-E}{KT}\right)}$$

Where, $n_v$ = Equilibrium number of vacancies
$N_t$ = Total number of atomic sites
E= Energy of formation of a vacancy
K= Boltzmann's constant
T= Absolute temperature

Usually, we can observe very large thermal energy near the melting temperature; there may be approximately as many as 1 vacancy per perfect 1000 atoms. Keeping in mind that this equation explains the equilibrium concentration of vacancies at a given temperature, quenching the material produced at high temperatures is one method to retain the equilibrium concentration of vacancies [3]. An example of a vacancy defect is the nitrogen-vacancy in diamond which is an important physical system for emerging quantum technologies, including quantum metrology, information processing, and communications.

## 3.2. Interstitial Defect

An interstitial defect occurs when a foreign atom occupies the crystal structure at a point that is normally unoccupied. In other words, when some constituent particles are positioned at an interstitial site, then that crystal is said to have interstitial defects. These types of defects are formed when a solute atom like an alloying or impurity element is positioned within a gap between the crystal lattice points of the solvent (base metal). Generally, the interstitial solute atoms are smaller than the solvent atoms; but they are larger than the interstitial site they occupy. As a result of which, the surrounding crystal structure gets distorted. Steel is an alloy of carbon and iron; here iron acts as a base metal and carbon acts as an interstitial atom. The size of the interstitial atom is about 50-85% less than that of the base metal atom. This increases the density of the substance. Interstitial defects are classified into 2 types which are as follows:

### 3.2.1. Self-Interstitial Defects

These are the type of interstitial defects that contain only those interstitial atoms which are already present in the crystal lattice. The structure of interstitial defects has been experimentally determined in some metals and semiconductors [4].

### 3.2.2. Impurity Interstitial Defects

These types of interstitial defects generally occur due to the positioning of the outside atoms into the interstitial site of the host crystal.

### 3.2.2.1. Effects of Interstitial Atoms on the Crystals [4]

- Interstitial atoms can alter the physical and chemical properties of materials.

- Interstitial carbon atoms have a significant role in altering the properties and processing of steel.

- Impurity interstitial atoms can be used for the storage of hydrogen in metals.

- The amorphization of semiconductors during ion irradiation is explained by the buildup of a high concentration of interstitial atoms leading to the collapse of the lattice as it becomes unstable [5, 6].

- The creation of large amounts of interstitial atoms in solid materials during mechanical alloying leads to high energy build up. These high-energy states can be released by annealing.

- It has been proposed that interstitial atoms are related to the onset of melting and the glass transition [4].

## 3.3. Substitutional Defects

A substitutional defect is formed when one atom or ion is replaced by a different type of atom or ion, occupying the normal lattice site [7]. If the substitutional atoms or ions are larger than the normal lattice site; then the surrounding interatomic spacings will be reduced and this causes the surrounding atoms to have larger inter-atomic spacings. These types of substitutional atoms or ions can be introduced either as an impurity or as an alloying material.

Examples of substitutional defects include the addition of phosphorus (P) or boron (B) into Si to prepare semiconductors. Similarly, in brass, the zinc atoms replace some of the copper atoms, which have a lesser radius compared to zinc. The substitutional atoms or ions greatly influence the mechanical, electrical, chemical, thermal properties of metallic materials. Substitutional defects can also present in ceramic materials. The occupancy of substitutional atoms or ions into either interstitial or substitutional sites mainly depends upon the size and the valence of substitutional atoms or ions. Fig. (**3**) represents the various point defects [7].

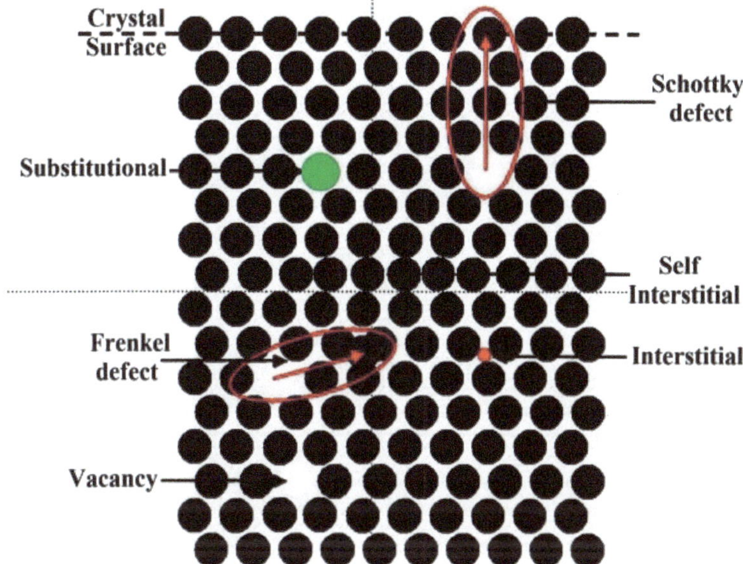


**Fig. (3).** Representation of various point defects [7].

In the case of ionic crystals, the point defects mainly depend upon the concept of charge neutrality and they are mainly classified as follows:

***Frenkel Defect:*** When an ion is displaced from its regular position creating a vacancy and the displaced ion moves to an interstitial position; the pair of the vacancy-interstitial defect together is called a Frenkel defect and is depicted in Fig. (**4**) [8]. Generally, the size of cations is much smaller than anions and hence displaced from the regular position and occupies the interstitial position.

***Schottky Defect:*** A pair of ions; one cation and one anion will be missing from an ionic crystal, without violating the condition of charge neutrality, when the valency of ions is equal. The pair of vacant sites, thus formed, is called a Schottky defect and is shown in Fig. (**5**) [8]. This type of point defect is dominant in alkali halides. These ion-pair vacancies like single vacancies facilitate atomic diffusion.

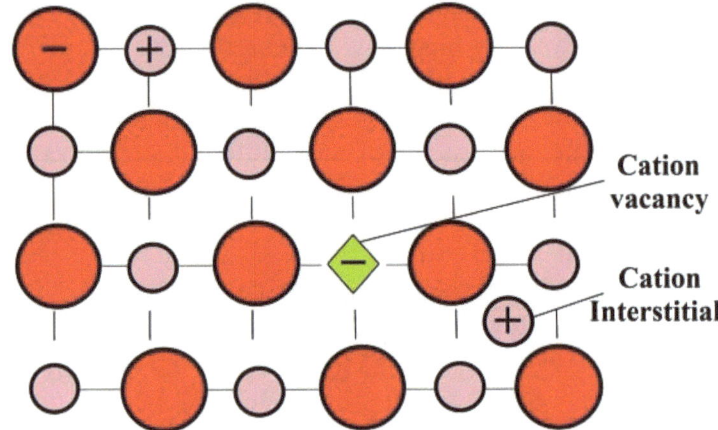


**Fig. (4).**  Frenkel defect [8].

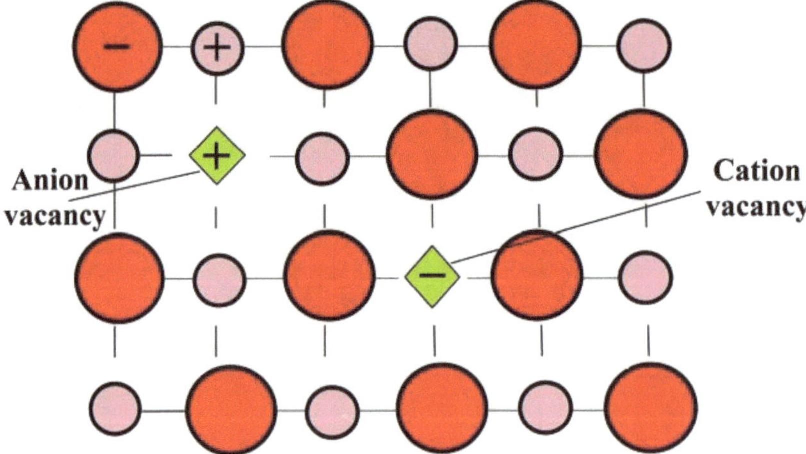


**Fig. (5).**  Schottky defect [8].

## 4. LINE DEFECTS

Line defects are another type of defect in crystals; where the row or line of atoms is deviated from their regular arrangement. Sometimes, an extra row or line of atoms is added and sometimes the same will be missing from the crystal structure. These types of defects are generated when stress is applied. The motion of defects follows the mechanism "slip" and finally results in plastic deformation. By using transmission electron microscope images of metals; many researchers had proved that the strength and ductility of metals are manipulated by dislocations. Generally, the dislocation is a boundary between the slipped region and the unslipped region.

There are mainly two types of dislocations, "edge dislocation" and the "screw dislocation". In reality, the edge and screw dislocations are just types of potential dislocations that can occur in metals.

## 4.1. Edge Dislocation

The edge dislocation is a type of line defect in which an extra half-plane of atoms is inserted in the regular crystal structure. If the insertion of an extra half-plane of atoms takes place from the top, then it is called positive dislocation. Similarly, if the insertion of an extra half-plane of atoms takes place from the bottom, then these types of defects are called negative dislocation. Fig. (**6**) represents the positive edge dislocation. The motion of edge dislocation is due to the applied shear stress.

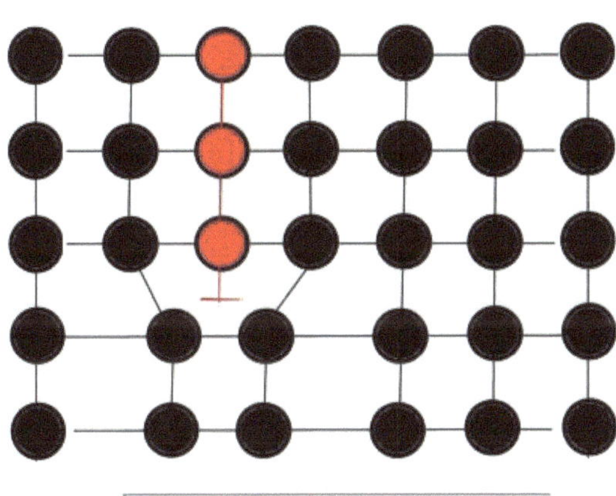

**Fig. (6).**  Positive edge dislocations.

## 4.2. Screw Dislocation

Screw dislocation is another type of line defect and it is quite difficult to visualize. The movement of a screw dislocation is due to the applied shear stress, but the defect line moves perpendicular to the direction of the stress and the atom displacement, rather than parallel. One can imagine the screw dislocation as follows; screw dislocation can be produced by cutting the crystal partway through a knife and then shearing the one part of the crystal parallel to the other cut. Fig.

(7) depicts the screw dislocation [9]. It can be seen from the figure that only a portion of the bonds is broken at any given time. As was the case with the edge dislocation, movement in this manner requires a much smaller force than breaking all the bonds across the middle plane simultaneously [9].

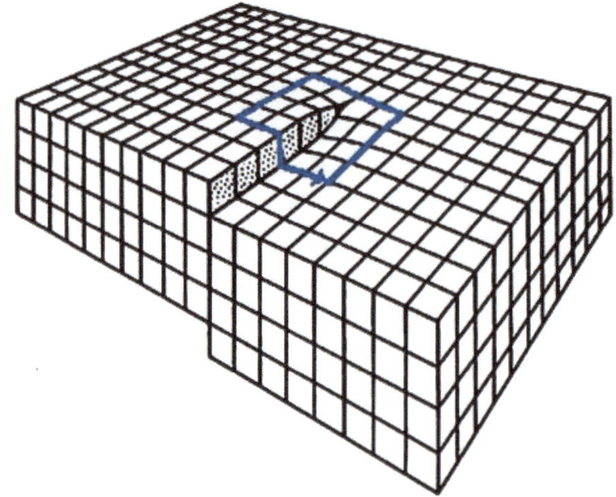

**Fig. (7).** Screw dislocation [9].

This dislocation transforms successive atomic planes into the surface of a helix around the dislocation line; hence it is named 'Screw dislocation'.

## 5. SURFACE DEFECTS

Surface defects are also called 'Planar defects'. They can be found in the boundaries or planes that separate material into regions. Each region having the same crystal structure but different orientations. Surface defects can cause corrosion and material failure.

The surface defects can significantly affect the corrosion resistance, mechanical properties, and surface properties of a material. There are different kinds of surface defects, but the important types are grain boundary defect, twin boundary defect, and stacking faults.

### 5.1. Grain Boundary Defect

Grain Boundary defect is a type of planar defect that separates regions of different crystalline orientations (grains) within a polycrystalline solid. Grain boundaries are usually formed due to the uneven growth of grains during crystallization. The grain boundary area generally depends upon the grain size of the crystalline solid, and it increases with a decrease in grain size. Fig. (**8**) shows the grain boundary defect [10].

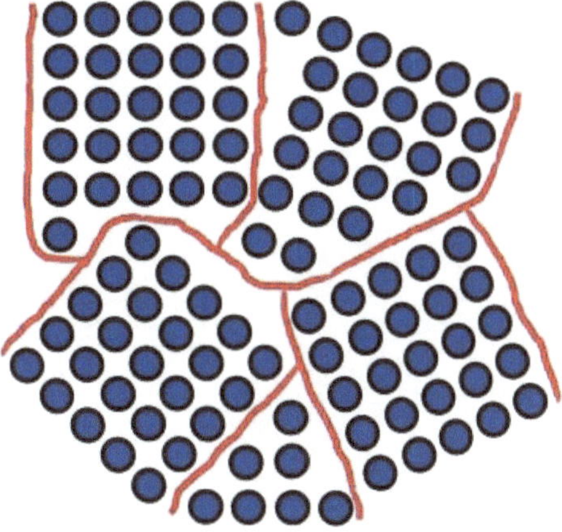

**Fig. (8).**  Grain boundary defect [10].

The lattice in each grain is identical but the lattices are oriented differently. The atoms are so close at some locations in the grain boundary that they cause a region of compression, and in other areas, they are so far apart that they cause a region of tension [10].

How to increase the strength of the material? Is it possible to control the grain growth?. The answer is YES. One method of improving the strength of the material is by controlling the grain size. By reducing the grain size, one can increase the number of grains, and in turn, the amount of grain boundary increases. More grain boundaries mean more strength; as all the dislocation moves only a short distance before encountering a grain boundary. The Hall-Petch equation relates the grain size to the yield strength:

$$\sigma_y = \sigma_o + \frac{K}{\sqrt{d}}$$

Where, $\sigma_y$ = Yield strength or stress at which the material permanently deforms
$\sigma_o$= Materials constant for the starting stress for dislocation
movement (or the resistance of the lattice to dislocation motion)
d= Average diameter of the grains
K= strengthening coefficient (a constant unique to each material)

## 5.2. Twin Boundary Defect

These are another type of surface defects and occurs at which the atomic arrangement on one side of the grain boundary is a mirror image of the atoms on the other side of the grain.

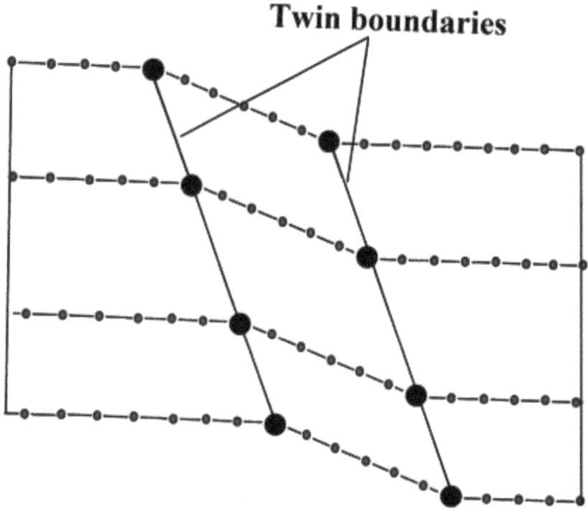


**Fig. (9).**  Pair of twin boundaries [11].

In other words; a twin boundary is a boundary that occurs when the crystal lattice on either side of a plane are mirror images to each other [11]. The boundary between the twinned crystals will be a single plane of atoms. The volume of the material whose orientation is the same as the mirror image of the matrix orientation is called as 'twin' and the mirror plane are called 'twinning planes'. Generally, twinning can occur during deformation, annealing, solidification, *etc*. The twin boundary defect is represented in Fig. (**9**) [11].

## 5.3. Staking Fault Defect

Generally, the crystal structures are described by stacking identical planes of atoms or different planes of atoms but indefinite manner. For example, a simple cubic structure is composed by stacking the identical plane of atoms in the manner A A A A A. Similarly, body-centered as AB AB AB and face-centered cubic structures as ABC ABC ABC. A stacking fault is a type of surface defect which characterizes the disordering of crystallographic planes from their regular stacking. Fig. (**10**) represents the stacking fault defect.

A——————————————————  A——————————————————

B——————————————————  B——————————————————

A——————————————————

B——————————————————  B——————————————————

A——————————————————  A——————————————————

      **Actual sequence of plane of atoms**        **Sequence of plane of atoms with stacking fault**

**Fig. (10).** Stacking fault defect.

## 6. VOLUME DEFECTS

Volume defects occur on a larger scale than the other types of crystal defects. However, for the sake of completeness and since they do affect the movement of dislocations, a few of the more common bulk defects are mentioned. Voids are regions where a large number of atoms missing from the lattice. Generally, voids occur due to the trapping of air bubbles during the solidification of materials, and it is commonly called porosity. When a void occurs due to the shrinkage of material during solidification is called cavitation [12]. The bulk defect can also occur when impurity atoms clusters together to form a small region of a different phase called precipitates; which are physically homogeneous [12].

## 7. STRENGTHENING OF MATERIALS

The ability of a metal or an alloy to plastically deform depends on the movement of dislocations. Strength is a measure of how easily a metal can undergo plastic deformation. Therefore, if we restrict the movement of dislocation; we can increase the strength. On the other hand, if the dislocation motion is easy, then the strength of a material decreases; materials will become soft and undergo deformation very easily. There are different ways one can use to improve the strength of the materials. They are strain hardening, grain boundary strengthening, solid solution strengthening, precipitation hardening, dispersion strengthening.

### 7.1. Strain Hardening

Strain hardening is also called work hardening and it is defined as a phenomenon where ductile metals become stronger and harder when they undergo plastic deformation at a temperature well below its melting point. The rate of strain hardening decreases with increasing temperature. Therefore, the materials are strain hardened at low temperatures called cold working. Strain hardening takes place when the dislocation density increases with plastic deformation (cold work) due to multiplication. Then, the average distance between dislocations decreases, and hence dislocations start blocking the motion of one another. As a result of this, the strength of the materials increases significantly.

On the other hand, during strain hardening, even the physical properties of the materials are also changes in addition to mechanical properties. The physical properties such as density, electrical conductivity, thermal coefficient of expansion, chemical reactivity, and corrosion resistance will vary during strain hardening.

The effect of cold work can be reduced by heating the material to a suitable temperature called annealing temperature. This facilitates regenerating the original properties into the material. This regeneration process consists of 3 stages such as recovery, recrystallization, and grain growth. Generally, most of the industries perform both cycles of strain hardening and annealing alternatively to deform the metals to a large extent.

## 7.2. Grain Boundary Strengthening

Grain boundary strengthening is also called Hall-Petch strengthening and it is defined as a method of strengthening the materials by altering their average grain size.

In the case of a polycrystalline metal, the grain size has a significant influence on the mechanical properties. This is due to the different crystallographic orientations of grains and therefore results in the formation of grain boundaries. Generally, deformation will follow a slip mechanism, and hence grain boundaries act as an obstacle to dislocation motion due to the following two reasons [13].

1. Dislocation must change its direction of motion due to the different orientation of grains.

2. Discontinuity of slip planes from grain one to grain two.

The stress required to move a dislocation from one grain to another to plastically deform a material mainly depends upon the grain or crystallite size. A lower number of dislocations per grain results in a lower dislocation pressure building up at grain boundaries. This makes it more difficult for dislocations to move into adjacent grains [13]. This relationship is mainly based upon the Hall-Petch relationship and it can be mathematically represented as follows:

$$\sigma_y = \sigma_o + \frac{K}{\sqrt{d}}$$

Where, $\sigma_y$ = Yield strength or stress at which the material permanently deforms
$\sigma_o$ = Materials constant for the starting stress for dislocation movement (or the resistance of the lattice to dislocation motion)
d= Average diameter of the grains
K= strengthening coefficient (a constant unique to each material)

According to the above equation, the yield stress is inversely proportional to the grain size of materials. As the grain size decreases, the repulsion stress felt by a grain boundary dislocation also decreases, and therefore maximum applied stress is required to propagate dislocations through the material. Fig. (**11**) shows the grain boundary strengthening [13].

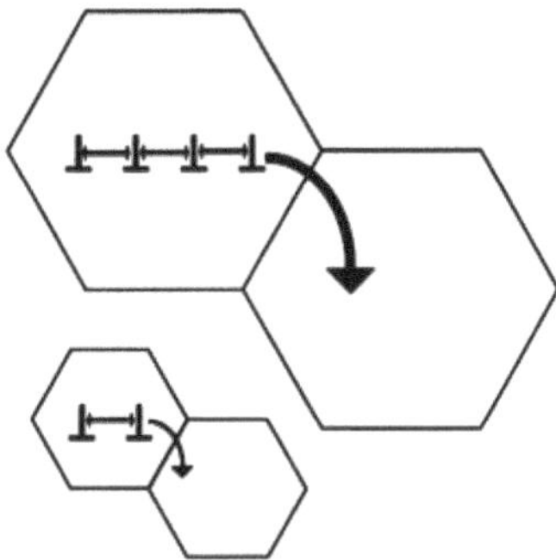

**Fig. (11).** Grain boundary strengthening mechanism [13].

The above figure depicts the concept of dislocation pile-up and how it affects the strength of a material. A material with a bigger grain size can have more dislocations pile up, leading to a larger driving force for the movement of dislocations from one grain to another. Therefore, very less force is required to move a dislocation from a larger than from a smaller grain, leading materials with smaller grains to exhibit higher yield stress [13].

## 7.3. Solid Solution Strengthening

Solid solution hardening is a process of dissolving one metal into another, similar to dissolving sugar into the water. It is an alloy in which the atoms of solute are distributed in the solvent and possess the same structure as that of the solvent. In other words, a solid solution will have the same composition but different structures. Always solute will be taken in a lesser amount and solvent will be in higher amount.

The solubility limit is a limit that the amount of solute can be dissolved into the solvent, after reaching the solubility limit solute remains undissolved in the solvent. For example, water can dissolve sugar up to a certain level, after the solubility limit all the excess sugar will remain undissolved and settles to the bottom. Further, the solubility limit can be increased by raising the temperature of the sugar solution, and then undissolved sugar will start dissolving in water. The same principle can be adopted even for metals during the preparation of an alloy.

Solid solution strengthening is of two types:

i. Substitutional solid solution strengthening

ii. Interstitial solid solution strengthening

### 7.3.1. Substitutional Solid Solution Strengthening

An example of substitutional solid solution strengthening is tin in copper to form bronze. In this case, the atoms of the solute material (tin) replace atoms of the solvent material (copper) in the crystal lattice as shown in Fig. (12) [14]. However, the solute and solvent atoms are having different sizes; they disturb the regularity of the crystal lattice. Then dislocations cannot easily move around this disturbance. It will take a much higher stress level or temperature to enable the dislocation to move again [14].

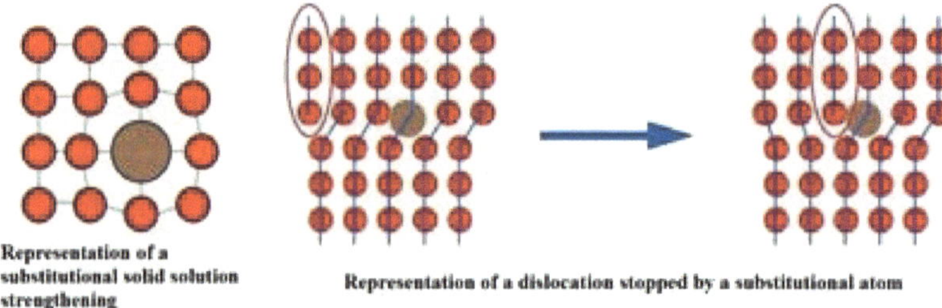


Fig. (12).  Substitutional solid solution strengthening [14].

### 7.3.2. Interstitial Solid Solution Strengthening

Another type of solid solution strengthening is an interstitial solid solution strengthening, and for example, is carbon in iron to form steel. In this case, solute atoms (carbon) are small enough to fit into the interstitial space between the solvent atoms (iron) in the crystal lattice as shown in Fig. (**13**) [14]. Once again, the alloying element catches the dislocation and prevents it from moving further. It then requires greater stress or thermal energy for the dislocation to move around the disturbing atom [14].

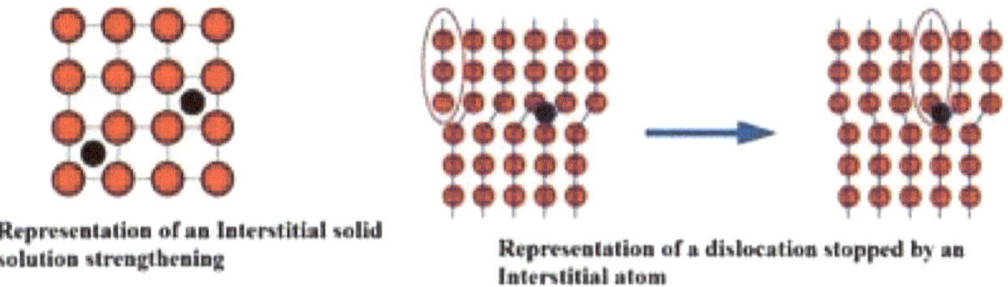


**Fig. (13).** Interstitial solid solution strengthening [14].


## 7.4. Precipitation Hardening

Another name for precipitation hardening is 'Age hardening'. It is a type of strengthening that involves heat treatment to impart strength to metals and their alloys. The name precipitation hardening is given due to the use of solid impurities or precipitates for the strengthening process.

The strength of materials can be amended by obstructing the motion of dislocations through the metals. This can be achieved by distributing the closely spaced sub-micron particles uniformly throughout an alloy. Then the particles; called precipitates, obstruct the dislocation motion through the alloy. Not all the alloys can be precipitation strengthened; the alloys such as Cu-Be, Al-Cu, 17-8 PH, and Al-Mg-Si can be strengthened.

The formation of these precipitates is accomplished by using a solution treatment at high temperatures before a rapid cooling process. The solution heat treatment results in a single-phase solution while the rapid cooling results in a stable material by preventing the creation and propagation of lattice defects. This greatly strengthens the metal matrix [15]. Generally, precipitation hardening is executed in a vacuum, inert atmosphere and at different temperatures ranging from between 900° and 1150° F. Even the processing time will also vary from one to four hours, depending on the exact material and the characteristics specified [15].


## 7.5. Dispersion Strengthening

It is the solid-state diffusion process in which very fine, submicron particles will be dispersed uniformly in the crystalline matrix to harden the alloy by hindering dislocation movements.

The dispersed particles may be insoluble particles during powder compaction or act as a precipitate in precipitation hardening. These second phase particles are proved to be significant materials in increasing the strength of materials. Most of the researchers are dispersing various second phase particles to improve the strength of engineering materials. The author of this book also published many research articles related to dispersion strengthening. Alloys based on iron, nickel, aluminum, and titanium can be incorporated with second phases like nanoparticles of zirconia, alumina, yttrium oxide, tungsten, *etc.* to improve the strength of the base metals. The mechanical properties of these types of alloys mainly depend upon the size, shape, and number of second phase particles. This method has allowed preparing high strength metal alloys.

## CONCLUSION

This chapter discussed the importance of defects in determining the strength of materials. Point defects, line defects, and surface defects are explained in a simple way. Frenkel and Schottky defects, edge and screw dislocations are discussed with suitable figures. This chapter also explains whether the defects or imperfections are strengthening friendly or not. We also discussed different strengthening mechanisms like strain hardening, grain boundary strengthening, solid solution strengthening, precipitation hardening, and dispersion strengthening.

## QUESTIONS

1. What are imperfections? Explain why do we need to study them?
2. What are the effects of imperfections over mechanical properties?
3. Mention various types of imperfections.
4. What are point defects and explain their various types briefly.
5. Differentiate Frenkel and Schottky defects.
6. What are line defects?
7. Explain the different types of line defects.
8. What are surface defects? Explain their different types.
9. What is the strengthening of materials? Discuss the various strengthening mechanisms.

## REFERENCES

[1]    J.W. Morris Jr, "Defects in crystals," In: Materials Science,    http://www.mse.berkeley.edu/groups/morris/MSE205/Extras/defects.pdf

[2]    askIITians, IIT JEE Chemistry, "Physical Chemistry, Solid State Imperfections in Solids and defects in Crystals",   https://www.askiitians.com/iit-jee-solid-state/imperfections-in-solids-and-defects-in-crystals/

[3]    D.R. Askeland, P.P. Fulay, and W.J. Wright, *The Science and Engineering of Materials.* 6th ed.

Cengage Learning: Stamford, USA, 2010.

[4]     Interstitial defect, Wikipedia, https://en.wikipedia.org/wiki/Interstitial_defect

[5]     D.N. Seidman, R.S. Averback, P.R. Okamoto, and A.C. Baily, "Amorphization processes in electron and/or ion-irradiated silicon", *Phys. Rev. Lett.,* vol. 58, no. 9, pp. 900-903, 1987. [http://dx.doi.org/10.1103/PhysRevLett.58.900] [PMID: 10035067]

[6]     G.F. Cerofilini, L. Meda, and C. Volpones, "A model for damage release in ion-implanted silicon", *J. Appl. Phys.,* vol. 63, no. 10, p. 4911, 1987. [http://dx.doi.org/10.1063/1.340432]

[7]     L. Morresi, Molecular Beam Epitaxy (MBE), CHAPTER 4, *Silicon Based Thin Film Solar Cells,* pp. 81-107, 2013.

[8]     "Point Defects in Ionic Crystals", © *H. Föll (Defects - Script).* https://www.tf.uni-kiel.de/matwis/amat/def_en/kap_2/illustr/t2_1_2.html

[9]     NTD Resource center, "Linear Defects – Dislocations", https://www.ndeed.org/EducationResources/CommunityCollege/Materials/Structure/linear_defects.htm

[10]    Engineering Archives, "Surface Defects", http://www.engineeringarchives.com/les_matsci_surface defects.html

[11]    S.V. Kailas, Chapter 3, "Imperfections in Solids, Material Science", https://nptel.ac.in/courses/Webcourse-contents/IISc-BANG/Material%20Science/pdf/Lecture_Notes /MLN_03.pdf

[12]    NTD Resource center, Bulk Defects, https://www.nde-ed.org/EducationResources/Community College/Materials/Structure/bulk_defects.htm

[13]    Wikipedia, "Strengthening mechanisms of materials", https://en.wikipedia.org/wiki/Strengthening_mechanisms_of_materials

[14]    M. Gedeon, Technical Tidbits, "Solid Solution Hardening & Strength", no. 16, April 2010.

[15]    R. Wojes, "Learn About Precipitation Hardening", https://www.thebalance.com/precipitation-hardening-2340019

CHAPTER 4

# Mechanical Properties of Materials

**Abstract:** The mechanical properties of a material reverberate the correlation between its response and deformation to an applied load. Some of the very important mechanical properties are ductility, strength, stiffness, and hardness. These properties can be studied by using various instruments available in the metallurgy or mechanical engineering laboratories. The mechanical properties of materials mainly depend upon the factors like nature of the applied load, its duration, and environmental conditions. The applied load may be tensile, compressive, or shear in nature; and its magnitude may be constant or may fluctuate continuously with time. Another important factor is time or duration, and it may vary from a fraction of a second to many years. Some of the important tests used to study mechanical properties are, creep test, tensile test, compression test, fatigue test, hardness test, impact tests, *etc.*

**Keywords:** Creep, Ductility, Elastic region, Engineering stress-strain curves, Fracture stress, Hardness, Malleability, Necking, Proof stress, Proportional stress, Strain, Strain hardening, Stress, Toughness, Tensile test, Ultimate tensile strength, Yielding region, Yield stress.

## 1. INTRODUCTION

Fig. (1) depicts the various mechanical properties of materials. Generally, the quality of a material can be investigated by studying their mechanical properties; then, materials will be rejected or accepted based upon the obtained results. Therefore, it is important to study the mechanical properties of materials. In this chapter, we have discussed various mechanical tests used to study the mechanical properties.

## 2. TENSILE TEST

Generally, the tensile test is used to investigate the strength, toughness, resilience, ductility and many other mechanical properties [1]. Solid materials will undergo deformation when they are subjected to load; and applied load can be tensile, compressive or shear.

Shashanka Rajendrachari & Orhan Uzun

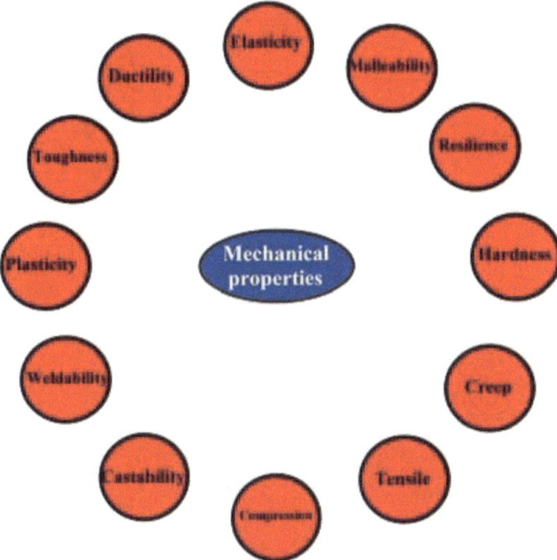


**Fig. (1).** Various mechanical properties of materials.

If the applied load is very less; then, the solid materials will maintain the same shape. Similarly, if you apply more load, then they will lose their shape. Fig. (**2**) represents the schematic diagram of a tensile testing machine [1].

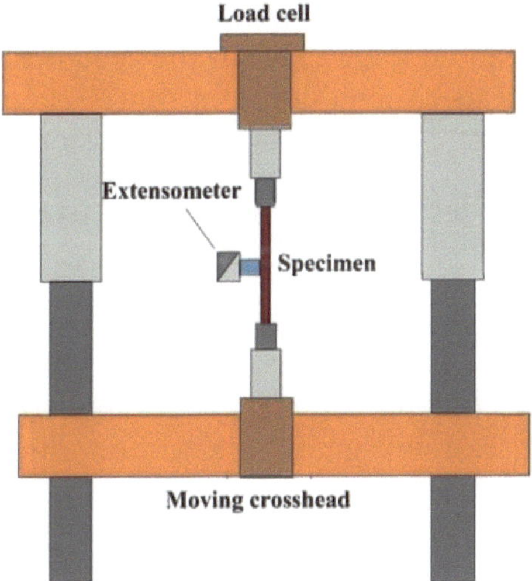


**Fig. (2).** Schematic diagram of a tensile testing machine [1].

The test specimens should have specific dimensions and shapes; it may be circular, square, rectangular cross-section [2], as shown in Fig. (**3**).

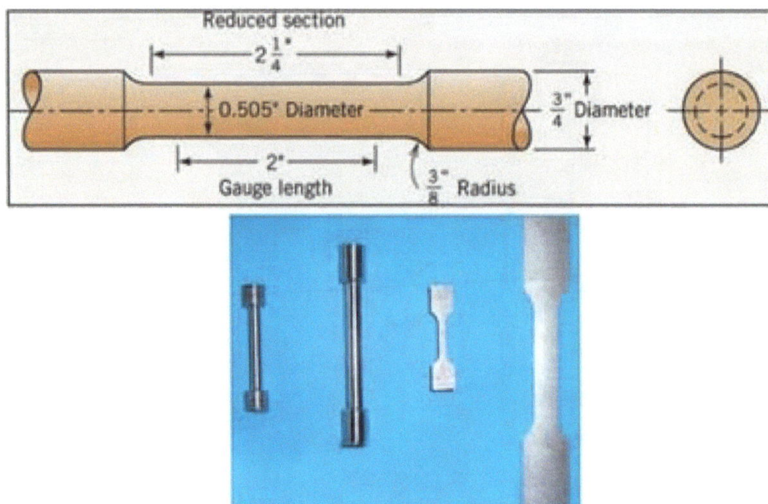


**Fig. (3).** The specimen for tensile test [1, 2].

While preparing the test specimens, care should be taken to avoid sharp changes in the section to reduce stress concentration. This brings down the chances of material failure at low-stress values. The specimen to be tested is firmly held between the two heads of the testing machine, as shown in Fig. (2). Then start increasing the tensile load continuously till the specimen breaks and note down the load recorded for the specimen during fracture. Generally, a Universal testing machine is used to perform tensile tests and compression tests.

**Gauge Length:** Tensile tests are performed on fixed lengths called "gauge lengths".

**Original Gauge Length:** Before conducting the tensile tests, two permanent marks are made on the test specimen at an appropriate distance called "Original gauge length". By subtracting the original gauge length (before the test) with final gauge length (after the test) will give the extent of tensile strength of the tested specimen.

## 3. ENGINEERING STRESS-STRAIN CURVES

The stress-strain curves can be obtained during testing a specimen. As we proceed with the test, the length of the specimen increases and the cross-sectional area decreases; but it is assumed here that dimension remains unchanged. Therefore, engineering stress and engineering strain can be explained by the following equations;

$$Engineering\ stress\ (\sigma_E) = \frac{Applied\ load\ (P)}{Original\ area\ of\ cross\ section\ (A_O)}$$

$$Engineering\ strain\ (e) = \frac{Change\ in\ length\ (dl)}{Original\ length\ (l_O)}$$

The load-bearing capacity of materials is called loadability. This is one of the important mechanical properties which decide the strength of materials. The loadability of materials can be studied by using a tensile testing machine (Universal testing machine) by plotting the engineering stress-strain curve. The engineering stress-strain curve is the same as that of the load-extension diagram. This is because, if the parameters on X-axes and Y-axes are divided by some constants, the original diagram remains unchanged. Therefore, only scales on the X and Y-axes will change, but the original diagram remains unaffected. If we divide the extension (X-axis) by the original gauge length, then the property becomes 'strain'. Similarly, if we divide the load (Y-axis) by the original area, then the property becomes 'stress'. Therefore, the load-extension diagram and engineering stress-strain diagrams are the same.

## 4. TRUE STRESS-STRAIN CURVES

During the tensile test, the specimen will be pulled apart slowly under applied load until it breaks. The specimen will behave differently from the start of the test until it breaks. These curves can be plotted by using the values of true stress and true strain generated instant to instant during the tensile test. Therefore, it will undergo different phenomena during the test, and it can be divided into four phases or regions, as illustrated in Fig. (**4**) [3]. Those four phases are as follows:

1. Elastic region
2. Yielding region
3. Strain hardening
4. Necking

The stress-strain curve shows how much the material can be stretched under the increasing stress [3]. As we proceed further, the length of the specimen increases with a decrease in the area of the specimen. Therefore, the value of true stress increases due to the gradual decrease in the cross-sectional area of the specimen. Similarly, the true strain decreases due to a gradual increase in length.

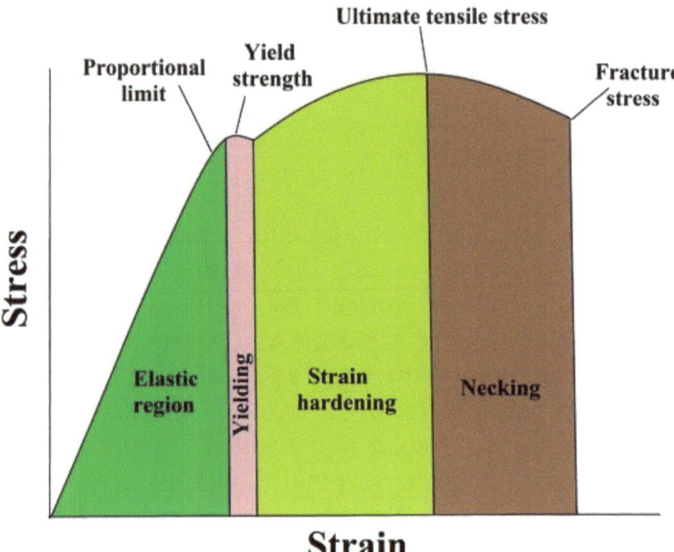


**Fig. (4).** Stress-strain diagram [3].

$$True\ stress\ (\sigma_T) = \frac{Instantaneous\ load\ (P_i)}{Actual\ cross-sectional\ area\ at\ that\ instant\ (A_i)}$$

True strain is generally denoted as ε, and it is the summation of all engineering strain from instant to instant.

$$True\ strain\ (\varepsilon) = \frac{l_1 - l_0}{l_0} + \frac{l_2 - l_1}{l_1} + \frac{l_3 - l_2}{l_2} \dots\dots$$

## 5. EVALUATION OF PROPERTIES

Tensile test is performed to evaluate the following mechanical properties, as discussed below. The properties are displayed in Fig. (4) stress-strain curve [3].

### 5.1. Proportional Limit or Stress

Proportional stress is the highest value of stress up to which the stress is directly proportional to strain, and their proportionality is linear, as shown in Fig. (4) [3]. Above this stress, the proportionality of stress and strain ends up, and the curve deviates from the linearity. Most of the time, the proportional limit is equal to an elastic limit for many metals.

## 5.2. Elastic Limit or Stress

Elastic limit or stress is defined as the highest value of stress up to which deformations are elastic in nature or temporary, and once the stress is removed, the materials will re-gain original shape. Above this stress, the materials will undergo permanent deformation or behaves like plastic. Stress and strain are not proportional to each other, but elastic stress is slightly more than the proportional stress. It is difficult to determine the exact elastic limit by stress-strain curves. To overcome this, the specimen should be applied with the load just above the proportional stress and unload the stress. The same procedure is followed several times with slightly increasing the applied load every time until the specimen does not show permanent deformation.

## 5.3. Ultimate Tensile Stress

Ultimate tensile strength is a measure of the highest value of stress that a material can withstand or sustain without fracture or breaking when it is stretched or pulled. Generally, the tensile property of a material is a very important phenomenon and it indicates how it will react to the forces applied [4]. Some of the materials break when a large amount of force is applied. On the other hand, some more metals will undergo elongation when force is applied. Materials that break very sharply when force is applied are called 'brittle materials'. Similarly, the materials which can be stretched or pulled over into wires during the applied load are called 'ductile materials'.

## 5.4. Breaking or Fracture Stress

Breaking stress or fracture stress is defined as the stress that material possesses at a point of breaking or failure. Breaking stress is calculated by using the formula;

$$Breaking\ stress = \frac{Force}{Area}$$

Therefore, breaking stress can also be defined as the force applied per unit area of the material. Breaking stress is one of the very important phenomena during the mechanical test; because it will give detailed information about how much metal or an alloy can be elongated to attain its ultimate tensile strength. It also depicts how much load a metal or an alloy can withstand before or during fracture.

## 5.5. Yield Stress

It is the stress at which material undergo appreciable plastic deformation at constant stress without any strain hardening. It can also be defined as the value of stress at yield point or yield strength.

Yield stress generally exists in materials like low carbon steel, mild steel, *etc*. Yield stress can be determined by using stress-strain curves as shown in Fig. (5) [4].

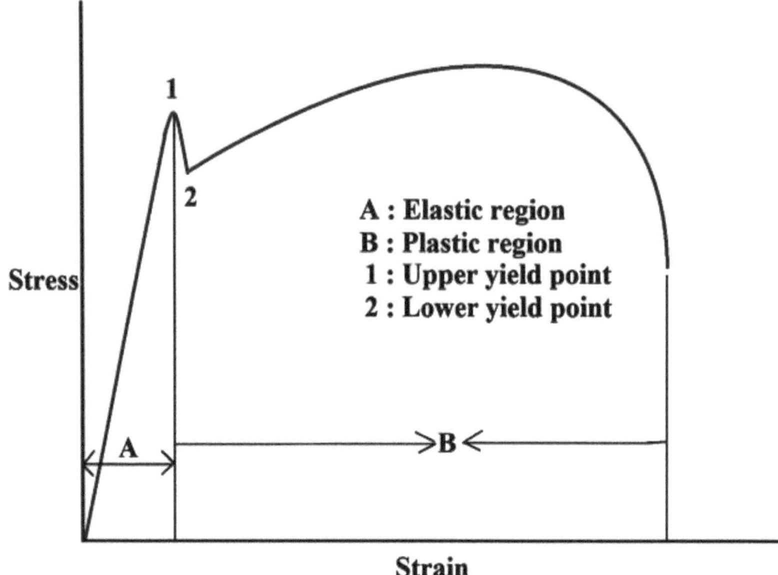


**Fig. (5).** Stress-strain curve showing the yield point [4].

When the material goes from the elastic region to the plastic region, a sharp break in the linearity can be observed in Fig. (5). The stress with respect to point 1 is called upper yield stress and the stress corresponds to point 2 is called as lower yield stress. Usually, the upper yield stress is affected by various conditions, whereas lower yield stress remains unaffected. Therefore, always lower yield stress is used to design components than higher yield stress due to its high safety factor.

## 5.6. Proof Stress

Proof stress can be studied for those materials which do not show appreciable yield point in the stress-strain curve. The proof stress is almost equal to the yield stress, and it is defined as a level of stress at which a material can undergo plastic deformation. It is also defined as the point at which the material will undergo some amount of plastic deformation equal to 0.2 percent. Proof stress can be calculated by a method called an offset method, as shown in Fig. (**6**) [4].

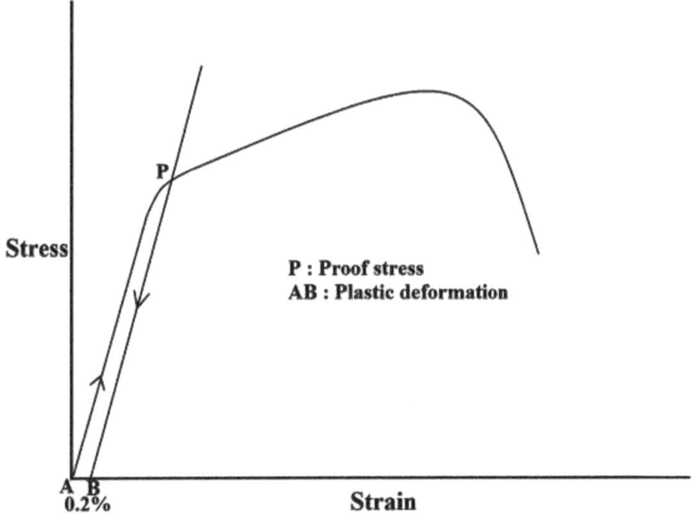


**Fig. (6).** Stress-strain curve with a method of calculating proof stress [4].

Some specified amount of plastic deformation is marked on the X-axis (strain axis). If the proof stress is to be calculated for 0.2% strain, then the point is marked on the stress-strain curve as 'B' as shown in Fig. (**6**) [4]. Plot a straight line from point B in such a way that should be parallel to the first linear line and should intersect the curve at point 'P'. The point 'P' corresponds to proof stress, as shown in the figure, and reports the value as proof stress. The specified amount of plastic deformation can be determined by applying the load to the specimen up to this point and unloads it.

## 5.7. Toughness

Toughness is the sum of elastic and plastic energies, and it is the total energy absorbed by a material before it breaks. Fig. (**7**) depicts the toughness in the stress-strain diagram [4]. Toughness is measured by calculating the area under the stress-strain curve from a tensile test and it is the unit of energy per volume. One of the metals having maximum toughness is Tungsten, and its toughness can be significantly increased by alloying with suitable metals.

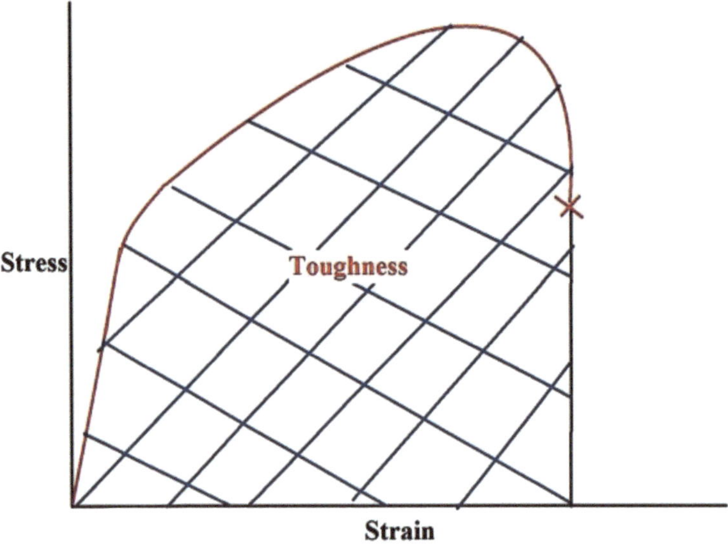



**Fig. (7).** The stress-strain curve showing the toughness [4].

## 5.8. Resilience

It is the ability of a material to rebound elastically into the original shape. In other words, it is the total energy absorbed by the materials during elastic deformation. Fig. (**8**) shows the resilience and modulus of resilience in load-extension and stress-strain diagrams, respectively [4]. It is the area up to the elastic limit, as shown in the figure.

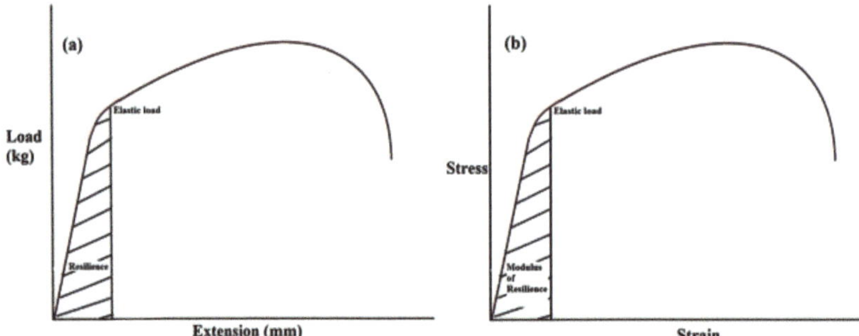



**Fig. (8). (a)** Load-extension diagram showing resilience. **(b)** Stress-strain diagram showing modulus of resilience [4].

The maximum amount of energy absorbed up to the elastic limit, without permanent distortion, is called proof resilience. Similarly, the modulus of resilience is defined as the maximum energy that can be absorbed per unit volume without permanent distortion. The unit of resilience is joule per cubic meter $(J/m^3)$.

## 5.9. Stiffness

Stiffness is expressed by Young's modulus, and it is defined as the resistance of material for elastic deformation.

$$Young's\ modulus = \frac{True\ stress\ (\sigma_T)}{True\ strain\ (\varepsilon)}$$

The values of both true stress and true strain are from the elastic region, and it does not affect by alloying or change in the microstructure.

## 5.10. Ductility

Ductility is an ability of a material to stretch under tensile stress before fracture. In other words, it can also be defined as an ability of a material to drawn in to thin wire. Generally, ductility can be expressed in terms of percentage elongation, and it is given as below;

$$\% \ Elongation = \frac{Final\ length\ (l_f) - Orginal\ length\ (l_o)}{Orginal\ length\ (l_o)} \times 100$$

Ductility of a material can be graphically represented as below:

Fig. (**9**) represents the load-extension diagram showing the ductility of materials [4]. Examples of ductile materials are copper, aluminum, steel, and some more metals. Ductility is a physical property, and it is not having any unit.

## 5.11. Malleability

Malleability is just opposite to that of ductility. In the case of the ductility test, tensile forces are used, but here compressive forces are used. It is defined as the ability of a material to undergo plastic deformation before fracturing under compressive stress. It can also be defined as the ability of a material to undergo rolling or flattening into thin sheets. Most of the metals with high ductility also possess greater malleability, and it can be measured by % reduction in the cross-sectional area of the material under study.

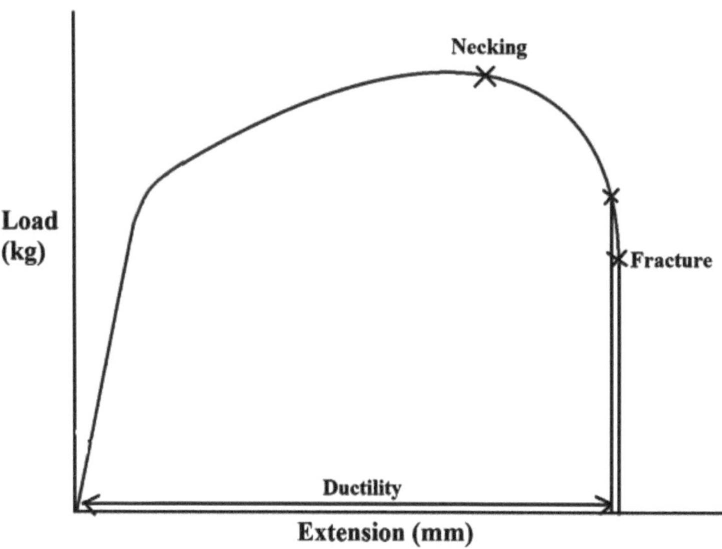


**Fig. (9).** Load-extension diagram showing ductility [4].

## 6. TYPES OF ENGINEERING STRESS-STRAIN CURVES FOR DIFFERENT TYPES OF MATERIALS

### 6.1. Ductile Materials without Yield Point

The engineering stress-strain diagram for ductile materials without a yield point is represented in Fig. (**10**) [4]. Due to strain hardening, the below curve shows the ascending trend of stress during plastic deformation. As we see from the figure, the curve starts dropping after the ultimate tensile stress because of the necking phenomenon. The distance between ultimate tensile stress and proof stress, as well as ultimate tensile stress and breaking stress, increases with an increase in the ductility of a material. Some of the metals and alloys which can undergo plastic deformation without yield point are nickel, copper, aluminum, austenitic stainless steel, silver, gold, platinum, *etc*.

### 6.2. Ductile Materials with a Yield Point

Fig. (**11**) represents the engineering stress-strain curve for ductile materials with a yield point [4]. The curve is almost similar to Fig. (**10**), but the only difference is the presence of a yield point near the elastic deformation region in Fig. (**11**), as shown. This yield point can be seen due to the formation of interstitial solid solution alloys having BCC and FCC crystal structures. Examples of interstitial elements are carbon, oxygen, nitrogen, boron, *etc.* Some of the metals and alloys which can undergo plastic deformation with yield point are mild steel, low carbon steels, oxide dispersed stainless steel (ODS), zinc, molybdenum, cadmium, *etc.*

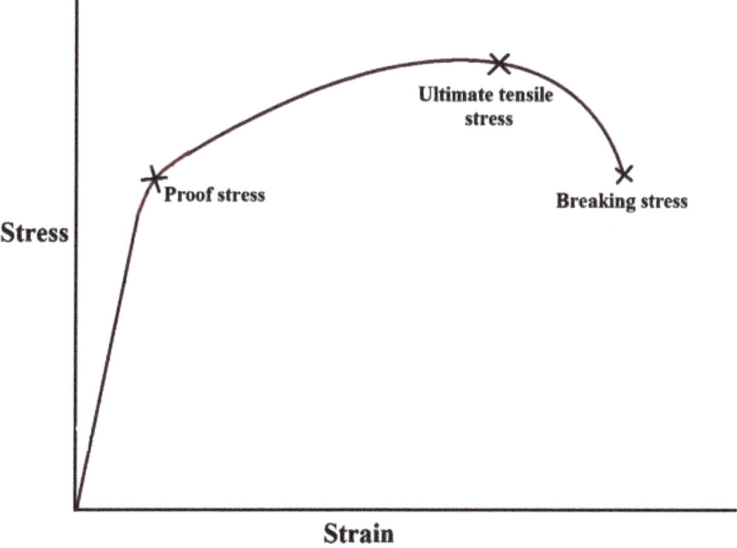


**Fig. (10).** Engineering stress-strain curve of ductile materials without yield point [4].

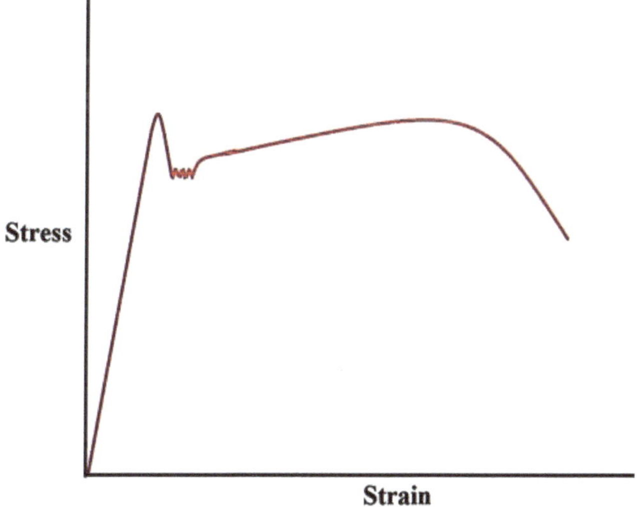


**Fig. (11).** Engineering stress-strain curve of ductile materials with a yield point [4].

## 6.3. Brittle Materials

Some of the brittle materials like gray and white cast iron, hardened steels, brass (equal proportions of Cu and Zn), concrete and carbon fibers do not show a yield point during plastic deformation and they do not undergo strain hardening. Hence, the distance between the ultimate tensile stress and breaking stress is negligible and sometimes have zero difference. In other words, ultimate tensile stress and breaking stress become the same. A typical engineering stress-strain curve for brittle material is represented in Fig. (**12**) [4].

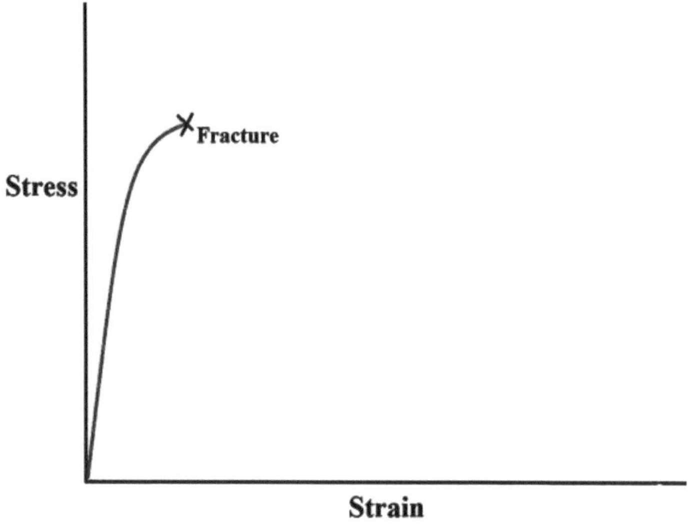


**Fig. (12).**  Engineering stress-strain curve for brittle material [4].

## 6.4. Polymeric Materials

In the case of polymers, the initial slope of the engineering stress-strain curve varies significantly from material to material. The graphical representation of the engineering stress-strain curve for polymeric materials is shown in Fig. (**13**) [4]. The curve is very much similar to that of ductile materials due to the crystallization of structure but not because of strain hardening. The polymers which show greater strain hardening are tougher and hence undergo ductile failure rather than brittle failure. Polymers behave as brittle materials at low temperatures and ductile materials at higher temperatures. As the temperature increases, one can observe a brittle to ductile transformation in polymers.

## 7. COMPRESSION TEST

Compression tests are performed to determine some of the fundamental mechanical properties. One can study how the material reacts when it is crushed or compressed by a specific load.

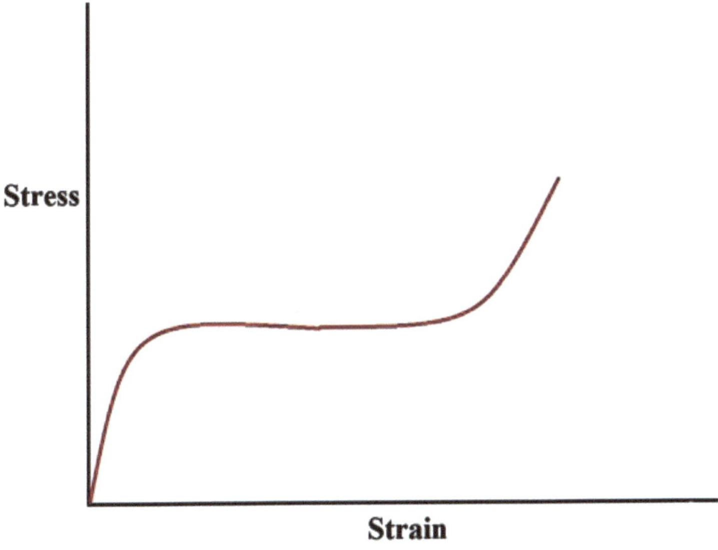


**Fig. (13).**  Engineering stress-strain curve for polymeric material [4].

Generally, a compression test is just opposite to that of the tensile test with respect to the direction of the applied load. Both compression test and tensile tests can be performed by using the same instrument 'Universal Testing Machine' (UTM). Brittle materials like ceramics, cast iron, and cement show good compression resistance than the tensile. Therefore, brittle materials will be used for the compression test to investigate their mechanical properties. Generally, brittle materials do not undergo plastic deformation under stress; therefore, the fracture stress is almost the same as ultimate tensile stress. We cannot find the necking phenomenon in the case of brittle materials as the fracture of the materials is clean and flat. The test specimen can be rectangular, square, spherical, but spherical cross-section specimens will be preferred to achieve uniform applied load. Fig. (14) shows the "Universal Testing Machine (UTM)" of Bartin University, Turkey used to study both tensile and compression tests.

**Fig. (14).** Universal Testing machine of Bartin University, Turkey.

## 8. FRACTURE

A fracture is defined as the separation of a material into two or more independent pieces under the action of stress (either tensile or fracture stress).

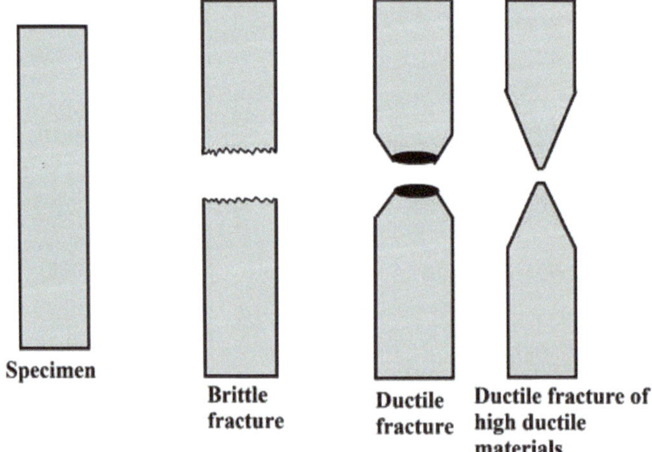


**Fig. (15).** Types of fracture during the tensile test.

Fig. (**15**) depicts the types of fracture that occur during a tensile test for brittle materials, ductile materials, and completely ductile materials.

## 9. CREEP

Creep is the material's property, and it is defined as the ability of a material to deform permanently under the influence of mechanical stress. Creep can take place due to the long-term exposure of materials to the high level of stress; even though the applied stress is less than its yield strength of the material [5]. The rate of deformation is called a creep rate and it is the slope of a line in creep strain *vs.* time curve. The rate of creep increases with an increase in temperature and it becomes more severe near their melting temperatures. At room temperature, creep is very less at a specific applied stress below the yield point. The rate of creep also depends upon the property of a material under test, time of exposure and amount of applied stress. Industries like nuclear power plants, furnace, boilers, jet engines, and heat exchangers are often victims of creep. There are mainly three stages of creep [6]; they are:

**Primary Creep:** It starts at an increased rate and slows down with time due to material hardening.

**Secondary Creep:** It has a relatively steady rate.

**Tertiary Creep:** It has an accelerated rate and ends when the material breaks. It is also associated with necking and formation of grain boundary voids.

Fig. (**16**) represents the graphical representation of creep behavior at the above-explained three stages [7].

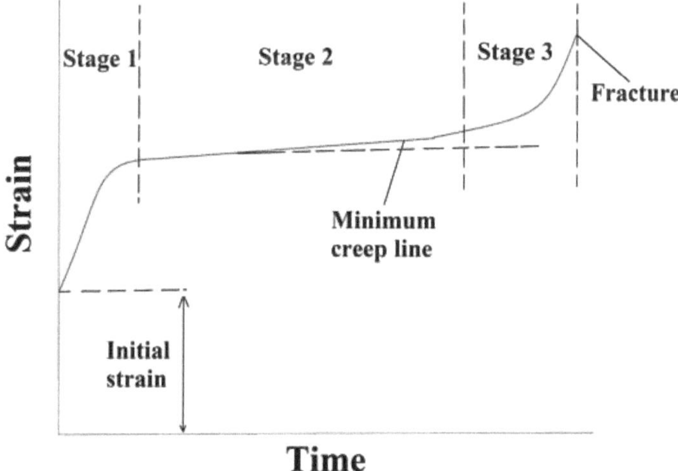


**Fig. (16).** Creep behavior at different stages [7].

The first stage of the creep is generally characterized by relatively fast plastic deformation and eventually, it decreases due to material hardening. During this stage, the creep resistance increases and results in a decrease of deformation rate.

The second stage of the creep occurs at a uniform and relatively low deformation rate. This deformation rate is a very important phenomenon significantly used to design the materials. The creep rate at this stage depends upon both the load (stress) and the temperature [7].

The third stage of the creep is associated with an accelerated strain rate caused by a decrease in the cross-sectional area of the specimen (necking). This stage is finalized by the specimen fracture [7].

Creep strength is another important physical quantity generally used to design the machines, instruments or structures working at higher temperatures.

## 10. CASTABILITY

Castability is also called 'fluidity' and it is defined as the ability of a molten metal to flow easily without premature solidification. It is one of the important properties of a metal because the poor castability of metals results in casting defects like incomplete filling in thinner sections of a mold. If the castability of metal is more, then it is easier to fill the thinner sections of the mold cavity with molten metal. This results in the production of a required shape of the metal mold with utmost precision. Fig. (**17**) shows the casting method used in Industries [8]. As the temperature increases, the castability of metal will also increase [8]. Therefore, elevated temperature always improves castability by decreasing the viscosity of molten metal. Whereas the impurities or inclusions present in the molten metal increases its viscosity and thereby decreases the castability.

**Fig. (17).** Casting method (Image credit: Shutterstock) [8].

Some of the factors that significantly influence the castability are as follows:

- Selection of materials: Select the materials having good castability.
- Shape of the mold metal: Design the mold at a low cost.
- Quality of the casted materials: It should be defect-free or having very few defects, specific dimensions, a good surface, and internal quality.

Some of the advantages of casting are:

1. Ease to manufacture.
2. Less cost for tooling.
3. Less energy is required.
4. Involving quicker development.
5. Very less waste.
6. Lower rejection rate.

## 11. HARDNESS

Hardness is a measure of resistance to localized plastic deformation caused by abrasion, indentation, scratching, cutting, or bending. Hardness mainly depends on ductility, elastic stiffness, plasticity, strain, strength, toughness, viscoelasticity, and viscosity [9].

Some metals, ceramics, and polymers are resistance to plastic deformation at the surface. On the other hand, elastomers and some polymers are resistant to elastic deformation at the surface [10]. One cannot give a specific fundamental definition to the hardness, because it is not a basic property of material, rather it is a combination of properties like yield strength, work hardening, true tensile strength, modulus, and other properties. Hardness measurements are widely used to define the quality of materials because they are quick, easy and nondestructive even though the marks or indentations are produced by the test, but they are in low-stress areas [10]. There are many methods used to determine the hardness of a substance, and few of them are explained below:

### 11.1. Mohs Hardness Test

German mineralogist Friedrich Mohs in the year 1812, devised a scale to test the hardness of a mineral called the Mohs scale for hardness measurement. It is one of the very old hardness measuring methods that use a scale ranging from 1 to 10 known as the Mohs hardness scale. The Mohs hardness test involves measuring the hardness of a mineral by calculating its ability to resist the abrasion or scratching by a substance of known hardness. To assign the numerical values to this physical property, minerals are ranked along the Mohs scale, as shown in Table **1**, which is composed of 10 minerals that have been given arbitrary hardness values [4, 10].

Mohs hardness test can give a significant result by identifying the minerals in the field based upon their hardness scale, but it cannot give accurate hardness for engineering materials like steel, ceramics, *etc*. Therefore, varieties of hardness measuring instruments were developed over the years to determine the hardness of engineering materials accurately. Hardness can be measured on the macro, micro, and nanoscale.

## 11.2. Brinell Hardness Test

The Brinell hardness test is one of the oldest hardness test methods still in use. This method was invented by Dr. J. A. Brinell from Sweden in the year 1900. The Brinell hardness test method is used to determine the hardness of those materials whose structure is very much coarse or having a rough surface. Generally, the hardness of casted and forged samples will be determined by the Brinell hardness method. Brinell hardness method often uses a very high applied load (3000 kgf) and a 10mm diameter indenter so that the resulting indentation averages out the most surface and sub-surface inconsistencies [11].

Table 1. Mhos hardness scale [4].

| Minerals | Mhos Hardness Scale |
|---|---|
| Talc | 1 |
| Gypsum | 2 |
| Calcite | 3 |
| Fluorite | 4 |
| Apatite | 5 |
| Orthoclase feldspar | 6 |
| Quartz | 7 |
| Topaz | 8 |
| Corundum | 9 |
| Diamond | 10 |

Once we apply a specific load to a ball indenter, then we will get Brinell hardness number, or simply the Brinell number. Brinell number is obtained by dividing the load used, in kilograms, by the measured surface area of the indentation in square millimeters left on the test surface [10]. Brinell testing is given in Fig. (**18**) [11].

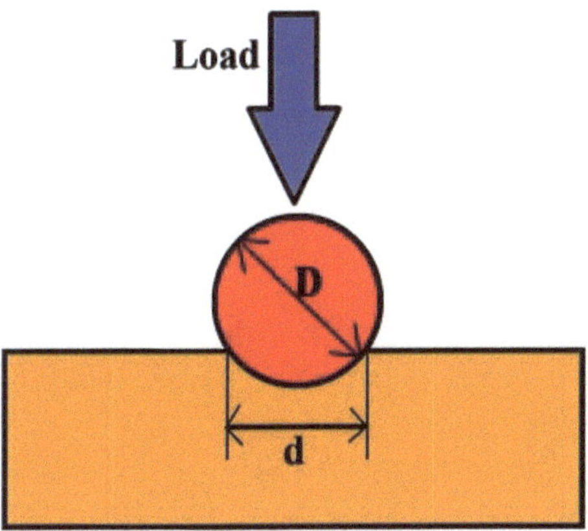


**Fig. (18).**  Brinell test [11].

Brinell hardness can be calculated by using the below formula as follows [11]:

$$HB = \frac{2L}{\frac{\pi D}{2}(D - \sqrt{D^2 - d^2})}$$

Where, HB = Brinell hardness (kgf/mm$^2$)
L = applied load in kilogram-force (kgf)
D = diameter of indenter (mm)
d = diameter of indentation (mm)

## 11.3. Rockwell Hardness Test

The Rockwell hardness test method is one of the most commonly used hardness test methods. It is easy to perform and gives accurate results when compared to other hardness methods. The Rockwell test is generally easier to perform, and more accurate than other types of hardness testing methods. The Rockwell method is used to measure the permanent depth of indentation produced by a specific load on an indenter. Initially, a test force or minor load is applied to a sample using a diamond or ball indenter. This preload breaks through the surface to reduce the effects of surface finish. After holding the preliminary test force for a specified dwell time, the baseline depth of indentation is measured [10]. The indenter may either made up of a steel ball of some specified diameter or a spherical diamond-tipped cone of 120° angle and 0.2 mm tip radius [12], as shown below in Fig. (**19**).

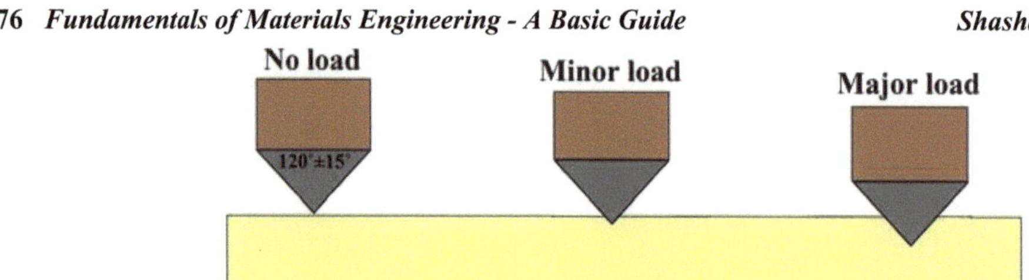


**Fig. (19).** Representation of minor load and major load by using diamond indenter [12].

During the test, a minor load of 10 kg is first applied, which causes a small initial indentation depth and this removes the effect of any surface irregularities. Then, the dial is set to zero and the major load is applied. Upon removal of the major load, the depth reading is taken while the minor load is still on. The hardness number may then be read directly from the scale. The indenter and the test load used to determine the hardness scale that is used (A, B, C, *etc.*) as shown below in Table **2** [10].

Table 2. The indenter and the test load used to determine the hardness scale that is used (A, B, C, *etc.*) [10].

| Scale | Examples |
|---|---|
| A | Cemented carbides, thin steel, and shallow case hardened steel |
| B | Copper alloys, soft steels, aluminum alloys, malleable iron, *etc.* |
| C | Steel, hard cast irons, pearlitic malleable iron, titanium, deep case hardened steel and other materials harder than B 100 |
| D | Thin steel and medium case hardened steel and pearlitic malleable iron |
| E | Cast iron, aluminum and magnesium alloys, bearing metals |
| F | Annealed copper alloys, thin soft sheet metals |
| G | Phosphor bronze, beryllium copper, malleable irons |
| H | Aluminum, zinc, lead |
| K, L, M, P, R, S, V | Bearing metals and other very soft or thin materials, including plastics. |

## 11.4. Vickers Microhardness Test

The Vickers hardness test was developed by Robert L. Smith and George E. Sandland in the year 1921 as an alternative to the Brinell hardness method [13]. This method is very easy and simple compared to all the hardness testing methods. Therefore, the Vickers hardness test is a more popular and widely used method to calculate both macro and microhardness.

The hardness calculations are independent of the size of the indenter and the indenter can be used for all types of materials irrespective of hardness. The unit of Vickers hardness is Vickers Pyramid Number (HV) or Diamond Pyramid Hardness (DPH) [14]. The hardness number is determined by the load over the surface area of the indentation, but not the area normal to the force and is, therefore, not pressure [14]. The Vickers microhardness tester is shown in Fig. (**20**). An applied testing load ranging from 10g to 1000g can be used. The use of a small amount of load results in small indentation that must be measured under a microscope. The measurements for hard materials must be taken at higher magnification (1000X) because the indents are very small at low magnification. The surface usually needs to be polished. The diagonals of the impression are measured and it is shown in Fig. (**21**).

**Fig. (20).**  Vickers microhardness testing machine at Bartin University, Turkey.

The Vickers microhardness value can be calculated using the relation [15]:

$$HV = 1.8544\,\frac{P}{d^2}$$

Where, P is the applied load and d is the diagonal length of the indentation.

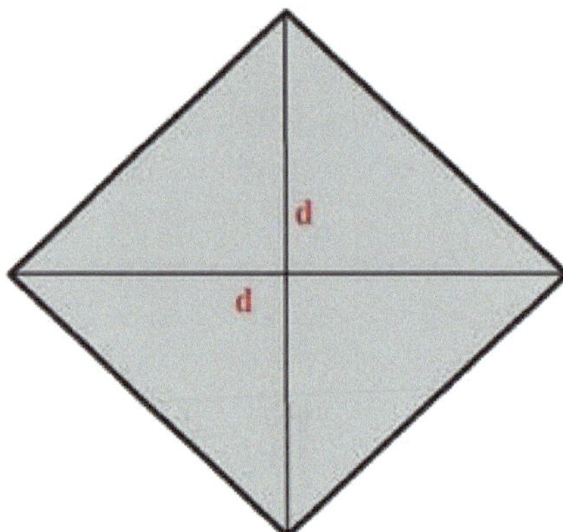


**Fig. (21).** Diagonals of the indentation.

## CONCLUSION

This chapter discussed some of the important terminologies used in the mechanical behavior of materials. Engineering and true stress-strain curves are explained in a very simple way. This chapter also contains important parameters used while evaluating the mechanical properties of materials. How polymers, ductile and brittle materials behave under stress are explained in detail. Different types of mechanical tests used to determine the mechanical properties are well discussed.

## QUESTIONS

1) Why we need to study mechanical properties?

2) Give some examples of mechanical properties.

3) Write a note on the tensile test.

4) What is gauge length and original gauge length?

5) Define engineering stress and engineering strain with mathematical equations.

6) What are true stress and true strain?

7) Explain briefly the true stress-strain curve with a neat labeled diagram.

8) Explain the following:

(i) Proportional limit (ii) Elastic limit (iii) Ultimate tensile stress (iv) Breaking stress (v) Yield stress (vi) Proof stress (vii) Toughness (viii) Resilience (ix) stiffness (x) Ductility (xi) Malleability

9) Explain briefly the types of engineering stress-strain curves for different types of materials:

(a) Ductile materials without yield point

(b) Ductile materials with a yield point

(c) Brittle materials

(d) Polymeric materials

10) Write a note compression test.

11) Define fractures and discuss the different types of fracture with diagrams.

12) Define creep and explain the different stages of creep.

13) What is castability? Mention the factors that significantly influence the castability.

14) Write some of the advantages of casting.

15) Define the term hardness and name some of the hardness methods.

16) Write a note on Mohs, Brinell, Rockwell and Vickers microhardness tests.

## REFERENCES

[1]     CAE WORLD, "Stress-Strain Diagram Part-1", *Aeronautical lecturing notes, .*

[2]     Wikipedia, "Tensile testing",  https://en.wikipedia.org/wiki/Tensile_testing

[3]     R.C. Hibbeler, "Mechanics of Materials",

[4]     V.D. Kodgire, and S.V. Kodgire, *Material Science and Metallurgy for Engineers.* 26th ed. Everest Publishing House, 2010.

[5]     Wikipedia, "Creep (deformation)",  https://en.wikipedia.org/wiki/Creep_(deformation)

[6]     Creep, "What is Creep? - Definition from Corrosionpedia", https://www.corrosionpedia.com/definition /344/creep-material-science

[7]     D. Kopeliovich, "Metal properties and tests", *SubsTech,* 2014.    http://www.substech.com/ dokuwiki/doku.php?id=creep

[8]     K. Than, "High-pressure key to lighter, stronger metal alloys, Stanford scientists find Stanford School of Earth, Energy &Amp", *Environ. Sci.,* no. May, p. 25, 2017.

[9]     Wikipedia, "Hardness",  https://en.wikipedia.org/wiki/Hardness

[10]    NDT Resource Center, NDT Course Material, "Materials and Processes, Hardness", https://www.nde-ed.org/EducationResources/CommunityCollege/Materials/Mechanical/Hardness.htm

[11]    Brinell Hardness Testing, "Newage hardness testing",  https://www.hardnesstesters.com/test-types/ brinell-hardness-testing

[12]    R. Shashanka, and D. Chaira, *Effect of Sintering Temperature and Atmosphere on Non lubricated Sliding Wear of Nano-Yttria Dispersed and Yttria-Free Duplex and Ferritic Stainless Steel Fabricated by Powder Metallurgy,* vol. 60, no. 2017, pp. 324-336, .

[13]    R.L. Smith, and G.E. Sandland, "An Accurate Method of Determining the Hardness of Metals, with Particular Reference to Those of a High Degree of Hardness", *Proc.- Inst. Mech. Eng.,* vol. I, pp. 623-641, 1922.
[http://dx.doi.org/10.1243/PIME_PROC_1922_102_033_02]

[14]    Wikipedia, "Vickers hardness test", https://en.wikipedia.org/wiki/Vickers_hardness_test

[15]    R. Shashanka, and D. Chaira, "Development of Nano-Structured Duplex and Ferritic Stainless Steels by Pulverisette Planetary Milling Followed by Pressureless Sintering", *Mater. Charact.,* vol. 99, pp. 220-229, 2014.

CHAPTER 5

---

# Polymers

**Abstract:** In the present chapter, we have discussed the classification of polymers, structure, properties, fabrication process, and polymerization mechanism. Generally, polymers are made up of a series of molecules joining together and the average molecular weight of chains ranges from 10,000 to more than one million. The process of chemically joining the monomers together to create giant molecules is called polymerization. In polymers, atoms are joined together by a strong bond called covalent bonding. Many of the polymers are organic (carbon-based polymers) and inorganic (non-carbon based polymers like polysiloxanes, polyphosphazene, polysilanes, *etc.*).

**Keywords:** Addition and condensation polymerization, Bakelite, Branched and cross-linking polymers, Classification, Homo and co-polymers, Linear, Mechanisms, Molding techniques, Natural polymers, Nylon, Polymers, Properties of polymers, Polyethylene, PVC, Poly propylene, Synthetic polymers, Tacticity, Thermosetting and thermoplastics.

## 1. INTRODUCTION

The word polymer comes from the Greek language "poly" meaning many and "mers" means units. A polymer is defined as a group of many units or it is a combination of a large number of monomers together to form a giant structure. Most of the time the word polymer is used for "plastic" also; but, all plastics are polymers, but not all polymers are plastics [2]. Fig. (**1**) shows the polymers [1].

Polyvinyl chloride (PVC), polyethylene, polymethyl methacrylate (PMMA), nylons, bakelite, *etc.* are some of the examples of polymers. Polymers are very popular due to their wide range of applications as mentioned below [1]:

- Automobile industries
- Constructional purposes
- Aerospace applications
- Medical applications
- Packaging applications
- Electronic goods *etc.*

**Shashanka Rajendrachari & Orhan Uzun**

**Fig. (1).** Polymers [1].

## 2. PROPERTIES OF A POLYMER

Polymers exhibit the following important properties:

1. Polymers are very resistant to chemicals.
2. Polymers are thermal and electrical insulators.
3. Lightweight, high strength-to-weight ratio, corrosion resistance.
4. Polymers can be processed in various ways.
5. Polymers can be molded in many colors, materials, and texture.
6. Properties of polymers can be easily modified by adding additives.
7. Polymers can be used to make items that have no alternatives from other materials.

## 3. CLASSIFICATION OF POLYMERS

Polymers are classified into many types based upon the source, structure, tacticity, properties, and monomeric units. Fig. (2) illustrates the classification of polymers [3].

### 3.1. Based on The Source [3]

Based on the source, the polymers are classified mainly as natural and synthetic polymers.

### 3.1.1. Natural Polymers

Natural polymers are those polymers that are present in natural sources like plants and animals. These are naturally occurring polymers.

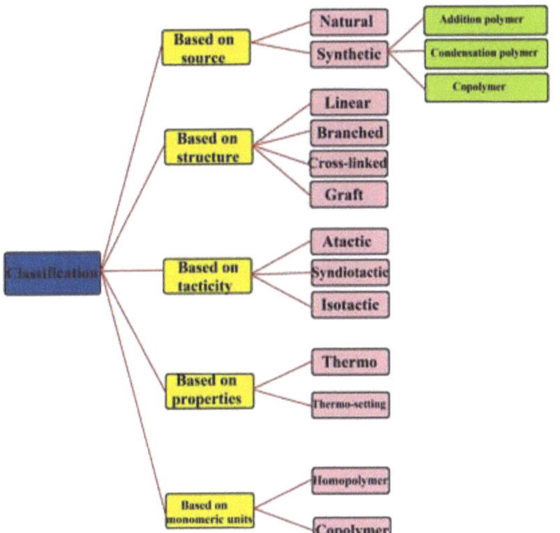

**Fig. (2).** Classification of polymers [3].

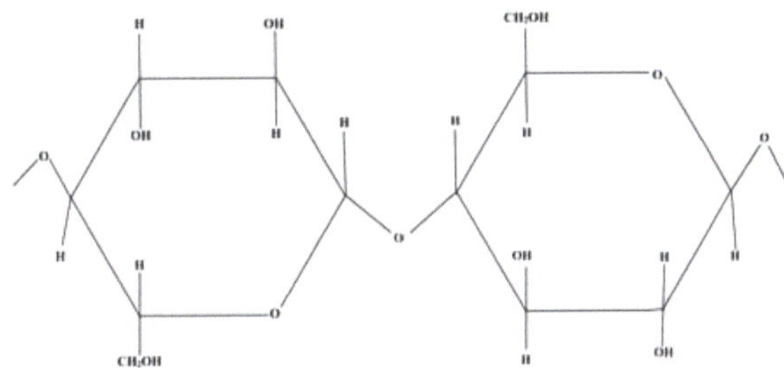

**Fig. (3).** Structure of cellulose [3].

Some of the important examples of natural polymers are proteins (found in animals), starch, rubber and cellulose (found in plants). The structure of cellulose is shown in Fig. (**3**) [3].

### 3.1.2. Synthetic Polymers

Synthetic polymers are those polymers which are artificially prepared by humans. These types of polymers are produced for commercial purposes in industries on a large scale to satisfy human needs. Some of the commercially produced synthetic polymers are polyethylene (packaging), Nylon Fibers (clothes, fishing nets, *etc.*), PVC (pipes and storage), *etc.* Fig. (**4**) shows the structure of PVC [3].

**Fig. (4).** Structure of PVC [3].

## 3.2. Based on Structure

Based on the structure, polymers are classified into linear, branched, crosslinked, and graft polymers.

### 3.2.1. Linear polymers

These are the types of polymers, in which the monomeric units are joined together to form long single straight chains. These polymeric chains are stacked together over one another to give a well-packed structure. Some of the examples for the linear chain polymers are polyester, high-density polythene, nylon, PVC, *etc.* Fig. (**5**) shows the structure of nylon [3].

**Nylon 6,6**

**Nylon 6**

**Fig. (5).** Structure of nylon [3].

### 3.2.2. Branched Polymers

These types of polymers are made up of joining the linear chains of monomers to the linear chain polymer. Branched polymers exhibit low melting point, low density, more compact and symmetrical molecular conformations. An example of branched polymers is low-density polythene as shown in Fig. (**6**) [4].

**Fig. (6).** Low-density polyethylene (LDPE) [4].

### 3.2.3. Cross-linked Polymers

Cross-linked polymers are generally composed of bifunctional and trifunctional monomers. These polymers exhibit a strong covalent bonding and therefore they are very hard and brittle. They are also called network polymers; where, monomers are linked together to form a three-dimensional network. Bakelite and melamine are important examples of cross-linked polymers. Fig. (7) depicts the structure of Bakelite [5].

### 3.2.4. Graft Polymers

Graft polymers are metameric copolymers with a linear backbone of one composite and randomly distributed branches of another composite. Fig. (8) depicts how grafted chains of species B are covalently bonded to polymer species A [7]. Although the side chains are structurally distinct from the main chain, the individual grafted chains may be homopolymer or copolymers [6, 7]. Graft polymers can be used as impact-resistant materials, as membranes to separate gases or liquids, hydrogels, drug deliverers, thermoplastic elastomers, and compatibilizers for the preparation of stable alloys [6]. The common example of a graft polymer is high impact polystyrene, which consists of a polystyrene backbone with polybutadiene grafted chains.

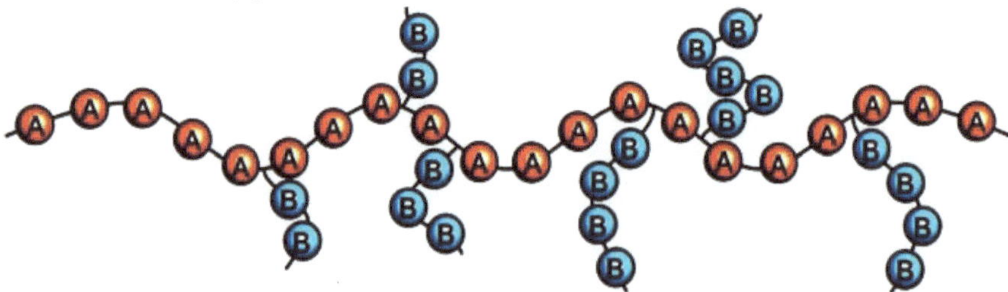

**Fig. (7).** Structure of Bakelite [5].

**Fig. (8).** Graft polymers [7].

## 3.3. Based on Tacticity

Tacticity is defined as an arrangement of substituents or adjacent chiral centers in a polymer either an ordered way or dis-ordered way. The tacticity helps in understanding some of the physical properties of a polymer like melting point, solubility, and also its mechanical properties [8].

### 3.3.1. Atactic Polymers

Atactic polymers are those in which the substituents or adjacent chiral centers are arranged randomly in a polymer. Polymers are formed by free-radical mechanisms such as polyvinyl chlorides; polystyrene is usually atactic in nature. Due to their irregular or random arrangement, atactic polymers are amorphous. In the presence of a suitable catalyst, it is possible to convert atactic polymers into syndiotactic polymers [8].

### *3.3.2. Syndiotactic Polymers*

In syndiotactic polymers, the substituents or adjacent chiral centers have alternate positions along the polymer chain. Gutta-percha is an example of a Syndiotactic polymer [9].

### *3.3.3. Isotactic Polymers*

Isotactic polymers are those polymers in which substituents or adjacent chiral centers are arranged on the same side of the polymer chain. Generally, they are semi-crystalline and form a helix configuration. Polypropylene formed by Ziegler-Natta catalyst is an isotactic polymer [10, 11]. Fig. (9) represents the atactic, syndiotactic, and isotactic arrangement in a polymer [10].

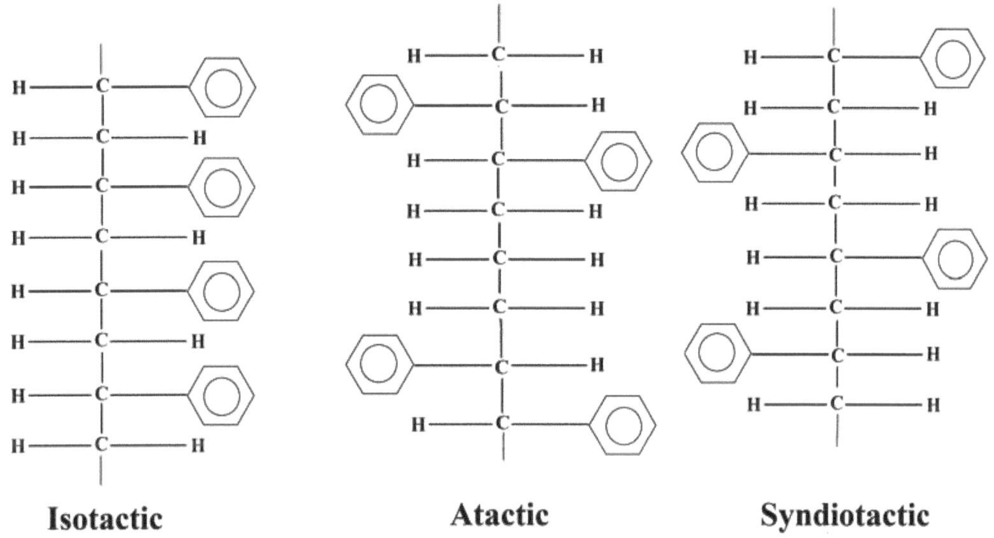

**Isotactic**             **Atactic**             **Syndiotactic**

**Fig. (9).** Representation of atactic, syndiotactic and isotactic polymer [10].

## 3.4. Based on Properties

### *3.4.1. Thermoplastic Polymers*

Thermoplastic polymers are plastic polymers that become soft on heating and hard on cooling. They can be heated and cooled several times without changing their chemical or mechanical properties [12]. During heating, the thermoplastic polymers melt to form a liquid. Some of the common examples of thermoplastics are polyvinyl chloride, polyethylene, nylon, polypropylene, polystyrene, *etc.*

### 3.4.2. Thermosetting Polymers

Thermosetting polymers are plastic polymer which becomes soft on heating and hard on cooling. As the reaction is irreversible, they cannot be remolded by reheating [12]. If we heat them, instead of becoming soft they will undergo burning or decomposition. Fig. (**10**) shows the representation of thermoplastic and thermosetting polymers [12]. Examples of thermosetting polymers are urea-formaldehyde, epoxy resin, phenol-formaldehyde (Bakelite), polyester resin. Most of the thermosetting polymers are hard, brittle, resistant to chemicals, thermal and electrical insulators.

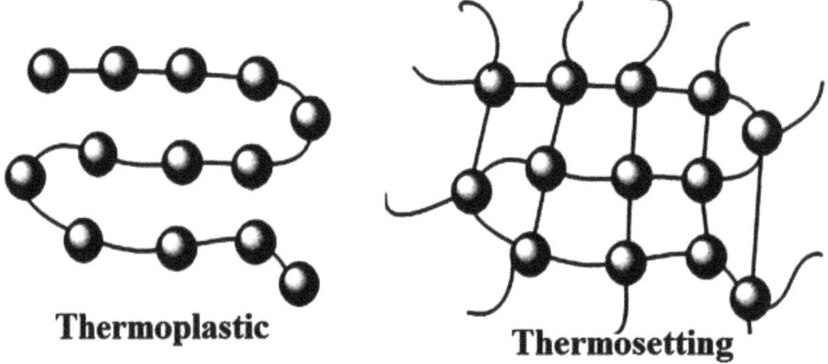


**Fig. (10).** Thermoplastic and thermosetting polymers [12].

**Table 1. Differences between thermoplastics and thermosetting polymers.**

| Thermoplastic Polymers | Thermosetting Polymers |
|---|---|
| These are formed by addition polymerization | These are formed by condensation polymerization |
| Monomer used is generally bifunctional | Monomer used is tri, tetra or poly-functional |
| They are long-chain linear polymer with negligible cross-links | These have three- dimensional network structure with the number of cross-links |
| They have low molecular weight | They have high molecular weight |
| They are soft, weak and less brittle | They are hard, strong and more brittle |
| They can be softened and reshaped and reused. | They cannot be softened and reshaped again. |
| Examples: polyethylene, polystyrene, PVC, PVA, *etc.* | Examples: phenol-formaldehyde, urea-formaldehyde, nylon 6:6, *etc.* |

## 3.5. Based on Monomeric Units

Based on monomeric units polymers are mainly classified into homopolymer and copolymers as represented in Fig. **(11)**.

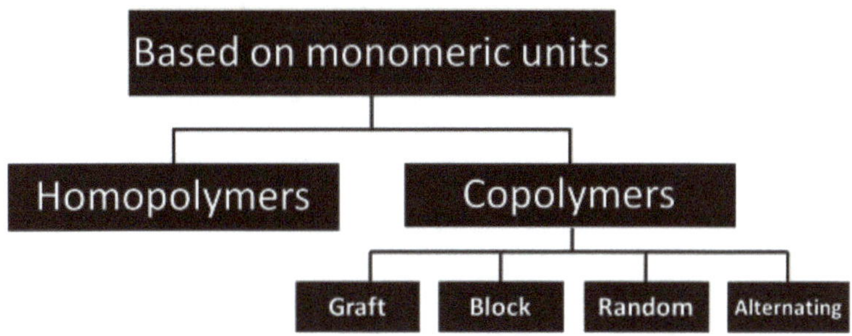

**Fig. (11).** Classification of polymers based on monomeric units.

### *3.5.1. Homopolymers*

A homopolymer is a polymer where the same types of monomers undergo polymerization and all the monomeric units in the polymer chain are the same as shown in Fig. **(12)**.

Examples: Polyvinyl chloride, polystyrene, polyethylene, *etc.*

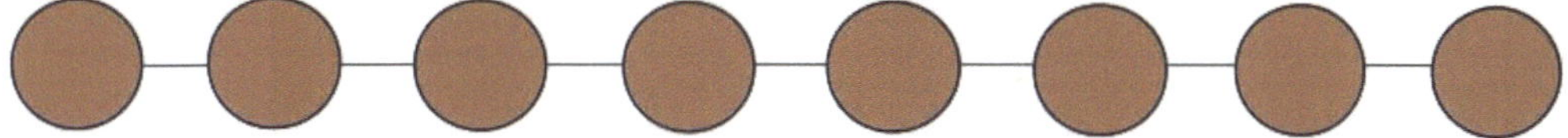

**Fig. (12).** Homopolymer.

### *3.5.2. Copolymers*

Copolymers are those when two or more different types of monomers undergo polymerization to form a polymer containing all the monomers used for polymerization. Fig. **(13)** shows the different types of copolymers.

Examples: Nylon 6,6, phenol-formaldehyde (Bakelite), urea-formaldehyde, *etc.*

## 4. TYPES OF POLYMERIZATION

The two important types of polymerization processes are shown in Fig. **(14)**.

## 4.1. Addition Polymerization

Addition polymerization is also called as "chain-growth polymerization". It is defined as a polymerization reaction in which monomers containing one or more double bonds are joined to each other without the elimination of any byproducts. Addition polymerization can usually take place in the presence of a free radical initiator. PVC, polyethylene, and polystyrene can be prepared by the addition polymerization method.

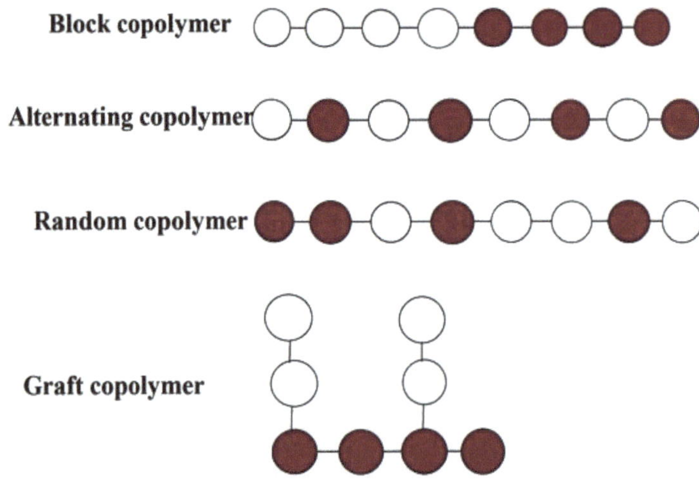

**Fig. (13).** Different types of copolymer.

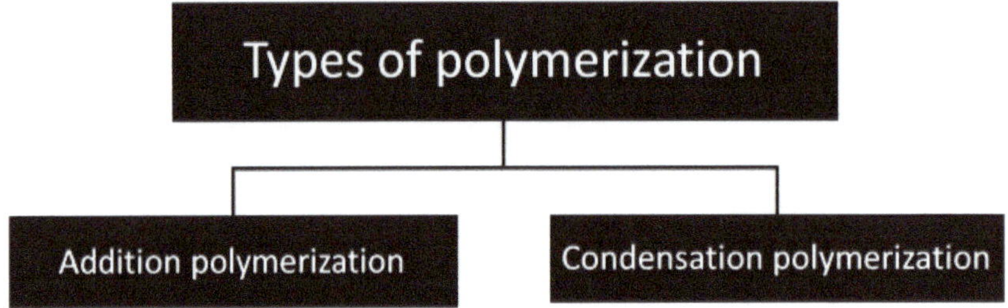

**Fig. (14).** Types of polymerization.

## 4.2. Condensation Polymerization

Condensation polymerization is also called as "step-growth polymerization". It is defined as a polymerization reaction in which bifunctional or polyfunctional monomers undergo an intermolecular condensation reaction with the elimination of byproducts like $H_2O$, HCl, $NH_3$, *etc.* Bakelite, polyester (Dacron) and polyamide (Nylon 6,6) can be prepared by condensation polymerization.

## 5. DIFFERENCES BETWEEN ADDITION AND CONDENSATION POLYMERIZATION

Some of the differences between addition and condensation polymerizations are given below [13].

**Table 2. Differences between addition and condensation polymerization.**

| Addition Polymerization | Condensation Polymerization |
| --- | --- |
| This polymerization reaction involves addition of monomers | This polymerization involves condensation reaction |
| It is a chain reaction of monomers | It precedes step wise |
| Monomers with double bonds undergo addition polymerization | Monomers with functional groups undergo condensation polymerization |
| No by-products will be eliminated | It involves elimination of by-products like $H_2O$ or HCl or $NH_3$ |
| The reaction is spontaneous (fast) | The rate of reaction is slow |
| Addition polymerization results in thermoplastics | Condensation polymerization results in thermosetting plastics |
| It is a chain growth polymerization | It is a step growth polymerization |
| Product yield does not depends upon duration of the reaction | Product yield depends upon duration of the reaction. More the reaction time, more is the product |

## 6. FREE RADICAL MECHANISM DURING ADDITION POLYMERIZATION [3]

Addition polymerization takes place through the free radical mechanism and it involves three different stages: initiation, propagation, and termination [14, 15].

### Initiation

A free radical is an atomic or molecular species having an odd or unpaired electron. They are highly active species.

(R·)

Benzoyl free radical

This part of initiation involves the addition of this radical to the first monomer molecule to produce the chain initiating species.

$$\dot{R} + CH_2 {=} CH_2 \longrightarrow R{-}CH_2{-}\dot{C}H_2$$

## Propagation

In propagation, the radical attacks another monomer to produce yet another free radical and the process continues until termination occurs.

$$R{-}CH_2{-}\dot{C}H_2 + CH_2{=}CH_2 \longrightarrow R{-}CH_2{-}CH_2{-}CH_2{-}\dot{C}H_2$$

$$\downarrow n CH_2 = CH_2$$

$$R{-}(CH_2{-}CH_2)_n{-}CH_2{-}\dot{C}H_2$$

## Termination

*Coupling or combination of two growing chains*

$$R{-}(CH_2{-}CH_2)_n{-}CH_2{-}\dot{C}H_2 + \dot{C}H_2{-}CH_2{-}(CH_2{-}CH_2)_n{-}R$$

$$\downarrow$$

$$R{-}(CH_2{-}CH_2)_n{-}CH_2{-}CH_2{-}CH_2{-}CH_2{-}(CH_2)_n{-}R$$

Dead polymer

*Coupling of growing chain with initiator free radical*

$$R{-}(CH_2{-}CH_2)_n{-}CH_2{-}\dot{C}H_2 + \dot{R} \longrightarrow R{-}(CH_2{-}CH_2)_n{-}CH_2{-}CH_2{-}R$$

Dead polymer

*Disproportionation*

$$R-(CH_2-CH_2)_n-CH_2-\overset{\bullet}{C}H_2 + \overset{\bullet}{C}H_2-CH_2-(CH_2-CH_2)_n-R$$

$$\downarrow$$

$$R-(CH_2-CH_2)_n-CH_2-CH_3 \quad + \quad CH_2\!=\!CH-(CH_2-CH_2)_n-R$$

Saturated dead polymer               Unsaturated dead polymer

# 7. IONIC POLYMERIZATION

Ionic polymerization is a chain-growth polymerization, but the active centers are ions or ion pairs [16]. The rate of polymerization in free radical polymerization generally depends upon the chemistry of monomers and stability of the radical used; whereas, the ionic polymerization mainly depends upon the reaction conditions. An impure monomer can ultimately lead to early termination and solvent polarity also has a great effect on the reaction rate. Typical solvents for ionic polymerization include non-polar molecules such as pentane, or moderately polar molecules such as chloroform [17]. There are mainly 2 types of ionic polymerization (i) Anionic polymerization, and (ii) Cationic polymerization.

## 7.1. Anionic Polymerization [18]

Similar to free radical polymerization; anionic addition polymerization takes place in the presence of initiators called anionic initiators like alkali metal amides (sodium or potassium amide) or alkali metal alkyls (n-butyl lithium). The monomers like styrene, acrylonitrile, and methyl methacrylate can undergo anionic polymerization. Anionic polymerization can proceed in 3 steps, (i) Initiation, (ii) Propagation, and (iii) Termination.

### *7.1.1. Initiation*

In the case of anionic polymerization, the initiator used is an anion (negatively charged ions). Different compounds can be used to generate anionic initiators but the most common compound is butyl lithium.

$CH_3-CH_2-CH_2-CH_2-Li$

**butyl lithium**

A bond between $CH_2$-Li will falls apart to form a positive lithium cation and a negative butyl anion called carbanion as shown below:

$$CH_3-CH_2-CH_2-CH_2-Li \longrightarrow CH_3-CH_2-CH_2-\overset{H}{\underset{H}{\overset{|}{\underset{|}{C}}}}\!\!{:}^{-} + Li^{+}$$

A pair of electrons from the butyl anion will be donated to one of the double bond carbon atoms of the monomer (ethylene). Now, this carbon atom already has eight electrons in its outer shell which it shares with the atoms to which it is bonded, so one pair of these electrons, specifically a pair in the carbon-carbon double bond, will leave the carbon atom, and settle on the other carbon atom of the carbon-carbon double bond. This forms a new carbanion, with the negative charge on that carbon as shown below. The process in which the butyl lithium undergoes bond breaking and the butyl anion reacts with a monomer molecule is called initiation.

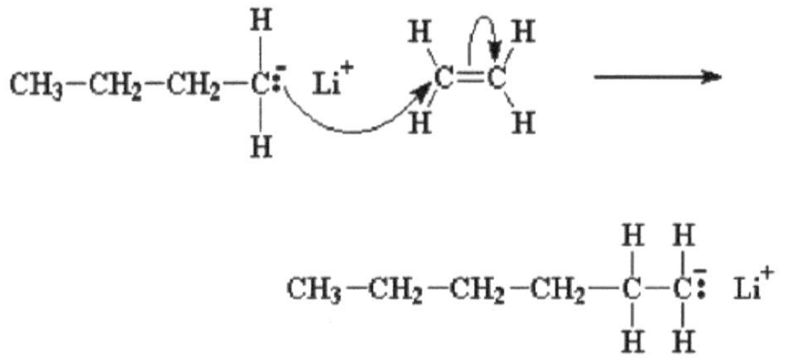

### 7.1.2. Propagation

The formed carbanion now reacts with another monomer (ethylene) molecule in the same way as the initiator reacted with the first monomer molecule in the initiation step and results in the formation of another carbanion. This reaction proceeds and every time another monomer is added to the growing chain; and a new anion is generated. In this way, the polymer chain starts growing and this process is called propagation.

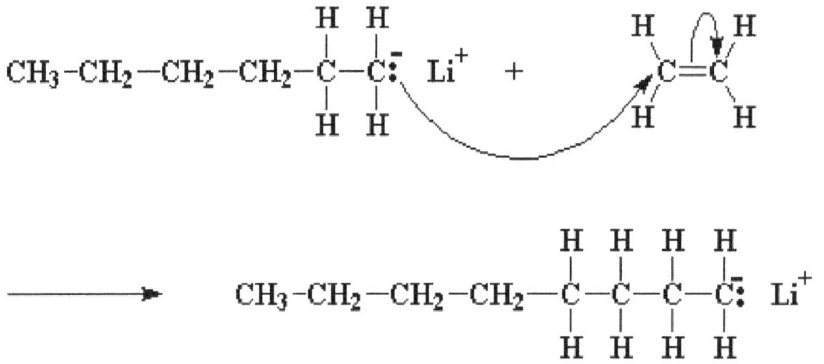

### 7.1.3. Termination

In this step, the growing polymer chain is terminated as follows:

$$CH_3-CH_2-CH_2-CH_2-\overset{\overset{H}{|}\ \overset{H}{|}\ \overset{H}{|}\ \overset{H}{|}}{\underset{\underset{H}{|}\ \underset{H}{|}\ \underset{H}{|}\ \underset{H}{|}}{C-C-C-\overset{..}{C}:}}\ Li^+ \xrightarrow{\ H\text{-}OH\ } \ \overset{\overset{H}{|}\ \overset{H}{|}}{\underset{\underset{H}{|}\ \underset{H}{|}}{+C-C+}_n}$$

## 7.2. Cationic Polymerization

In cationic addition polymerization, the initiator is a cation obtained either from protonic acids ($HCl$, $HNO_3$, $H_2SO_4$) or lewis acids ($AlCl_3$, $BF_3$, *etc.*).

**Initiation**

$$H^+ + CH_2{=}CH \longrightarrow CH_3-\overset{+}{C}H$$
$$\underset{R}{|} \qquad\qquad \underset{R}{|}$$
$$\text{Carbonium ion}$$

$$CH_3-\overset{+}{C}H + CH_2{=}CH \longrightarrow CH_3-CH-CH_2-\overset{+}{C}H$$
$$\underset{R}{|} \qquad \underset{R}{|} \qquad\qquad \underset{R}{|} \qquad\qquad \underset{R}{|}$$

$$\downarrow \ \text{Chain propagation}$$

$$CH_3-CH{+}CH_2-CH_2{)_n}\ CH_2-\overset{+}{C}H$$
$$\underset{R}{|} \qquad\qquad \underset{R}{|} \qquad\qquad \underset{R}{|}$$

**Propagation**

**Termination**

$$CH_3-CH{+}CH_2-CH{)_n}\ CH_2-\overset{+}{C}H \xrightarrow{-H^+} CH_3-CH{+}CH_2-CH{)_n}CH_2{=}CHR$$
$$\underset{R}{|} \qquad \underset{R}{|} \qquad \underset{R}{|} \qquad\qquad \underset{R}{|} \qquad \underset{R}{|}$$

## 8. SYNTHESIS OF SOME INDUSTRIALLY IMPORTANT POLYMERS

The first synthetic polymer phenol-formaldehyde resin (Bakelite) was discovered by Hendrik Baekeland (1863–1944) in the year 1907. During World war I in Germany, the first synthetic rubber known as methyl rubber was produced from 2,3-dimethyl butadiene as a potential substitute for natural rubber. But today we can prepare thousands of different types of polymers, plastics, rubbers, *etc.* Some of the important polymers used more often are polyethylene, polyvinyl chloride, polypropylene, polystyrene, polyamides, and bakelite. This part describes the synthesis, properties, and applications of a few industrially important polymers.

## 9. POLYETHYLENE

Polyethylene is a light synthetic resin exhibit a wide range of applications and it is a member of polyolefin resins. They are prepared from the polymerization of ethylene in the presence of an initiator. Fig. (**15**) shows the polyethylene beads [18].

**Fig. (15).** Polyethylene beads [18].

It is one of the very popular polymers in the world due to its wide range of applications. They are mainly used to prepare transparent food wrap, carry-bags, plastic bottles, and automobile fuel tanks. There are mainly two types of polyethylene as follows:

### 9.1. Low-density Polyethylene

It is generally prepared by the polymerization of ethylene gas under high pressure of 1000 to 2000 atmospheres at a temperature of 350 K to 570 K in the presence of a low concentration of a peroxide initiator and it follows free radical addition polymerization. Low-density polythene is flexible, tough, chemically inert and a poor conductor of electricity. Hence, it is used as an insulating wire, manufacturing of toys, flexible pipes, squeeze bottles, *etc.*

## 9.2. High-density Polyethylene

High-density polyethylene is formed during addition polymerization of ethylene in a hydrocarbon solvent in presence of a catalyst such as triethylaluminium and titanium tetrachloride (Ziegler-Natta catalyst) at a temperature of 333 K to 343 K at a low pressure of 6-7 atmospheres. Generally, high-density polyethylene consists of linear molecules and exhibit high density due to close packing. The properties like chemically inert and more toughness made them preferable to be used in the manufacturing of dustbins, buckets, bottles, pipes, *etc*.

$$-CH_2-\underset{\underset{\underset{CH_3}{|}}{\underset{CH_2}{|}}}{CH}-CH_2-CH_2-CH_2-CH_2-CH_2-\underset{\underset{\underset{CH_3}{|}}{\underset{CH_2}{|}}}{CH}-CH_2-$$

High density polyethylene

## 9.3. Properties of Polyethylene

- White, rigid and waxy
- Low density but high strength
- Exhibit maximum stiffness
- Thermal resistance
- Chemically inert
- High refractive index
- Appreciable stretchability

## 9.4. Applications of Polyethylene

- Bottles and bottle caps
- Kitchen and domestic appliances
- Toys
- Packaging
- Tubes

## 10. POLYVINYL CHLORIDE (PVC)

It is a hard and stiff synthetic amorphous solid (Fig. **16**) prepared from the polymerization of vinyl chloride [19]. It is the second popular polymer after polyethylene due to its huge production and applications. Its applications range from domestic products to industrial products. PVC is prepared by the addition polymerization of vinyl chloride in the presence of benzoyl peroxide and polymerization follows free radical mechanism.

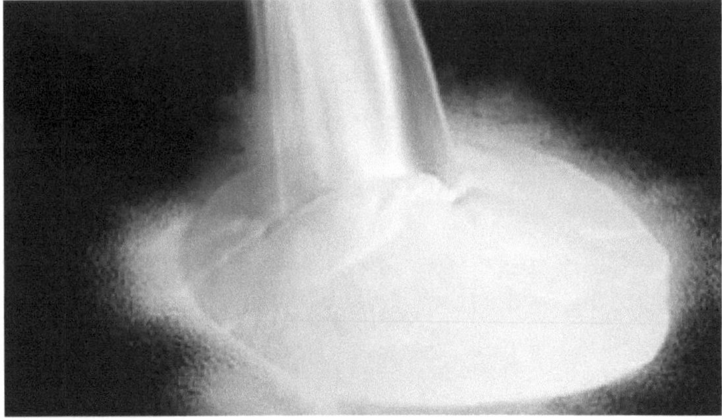

**Fig. (16).** Polyvinyl chloride powder [19].



## 10.1. Properties of Polyvinyl Chloride

- Hard and stiff amorphous solid
- Good flame resistant property
- Excellent strength
- Good Chemical inertness
- Thermal resistance
- Excellent blending nature

## 10.2. Applications of Polyvinyl Chloride

PVC is used to manufacture sheets, pipes, safety helmets, refrigerator components, tires, bicycle and motorcycle mudguards, raincoats, curtains, cable insulators, radio components, *etc.*

## 11. POLYPROPYLENE

Polypropylene is a type of polyolefin readily formed by polymerization of propylene with a suitable catalyst like aluminum alkyl and titanium tetrachloride (Ziegler-Natta catalyst). Fig. (**17**) depicts the polypropylene pellets [20]. The properties of polypropylene mainly depend upon the molecular weight, method of production, and the copolymers used. Polypropylene shows some advantages over polyethylene in improved strength, rigidity, and high-temperature stability.

**Fig. (17).** Polypropylene pellets [20].

Propylene            Polypropylene

## 11.1. Properties of Polypropylene

- Isotactic polypropylene has a density of around 0.90 g/cm$^3$
- It melts in the temperature range 165–170°C
- The electrical properties of polypropylene are similar to polyethylene
- It is inert to water and microorganisms
- It is a low-cost polymer

## 11.2. Applications of Polypropylene

Polypropylene is mainly used in the fabricating filaments and fibers, automotive and appliance components, packaging containers, furniture and toys, special devices like living hinges, *etc.* [21].

## 12. POLYSTYRENE

Polystyrene is a synthetic polymer prepared by free radical polymerization of aromatic hydrocarbon styrene as depicted below. Polystyrene was first discovered by Eduard Simon in the year 1839, an apothecary from Berlin [22].

## 12.1. Properties of Polystyrene

- Amorphous
- Transparent
- Hard
- Inexpensive resin per unit weight.
- Light Stable
- Moisture-resistant
- Exhibit poor oxygen and UV resistance
- More brittle
- Poor impact strength due to the stiffness of the polymer backbone

## 12.2. Applications of Polystyrene

Polystyrene is used in packaging, toys, combs, buttons, radio and television parts, battery cases, lenses, audio cassettes, to produce various foamed products like disposable drinking cups, egg cartons, trays, fast-food containers, and cushioned packaging.

## 13. POLYAMIDES [3]

Polyamides are synthetic polymers containing repeated amide groups (-CO-NH-) Some of the examples of polyamides are proteins (silk and wool), nylons (nylon-6, nylon-6,6, and nylon-6,10), aramids, *etc.* The number in the nylon is based on the number of carbon atoms in the repeating unit.

## 13.1. Nylon-6,6

Nylon-6,6 is obtained by polycondensation of equivalent amounts of hexamethylene diamine and adipic acid with water in a reactor. Then, nylon-6,6 salt containing an ammonium/carboxylate mixture crystallizes out. The nylon salt goes into a reaction chamber where the polymerization process takes place either in batches or continuously [23]. The chemical reaction of hexamethylene diamine and adipic acid to form nylon-6,6 is given below:

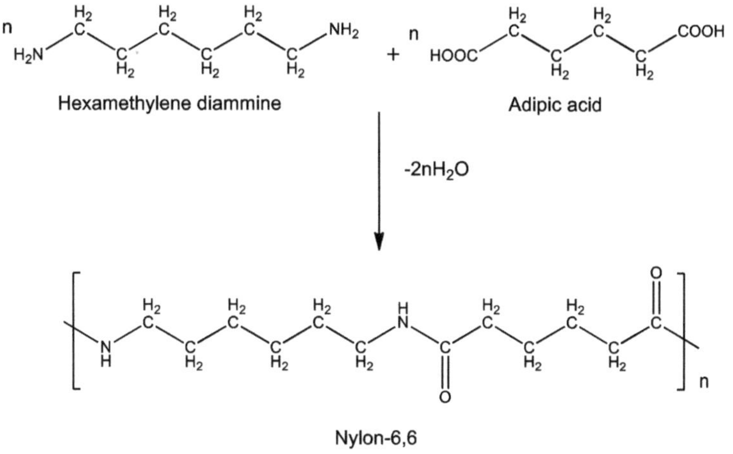

### 13.1.1. Properties

- It exhibits high mechanical strength, hardness, and toughness
- Good mechanical damping ability
- Good sliding properties
- High fatigue resistance
- Good thermal stability
- Excellent wear resistance

### 13.1.2. Applications

Nylon-6,6 is mainly used as fibers for textiles and carpets. It is also used in airbags, apparel, radiator end tanks, rocker covers, air intake manifolds, oil pans, and ball bearing cages.

## 13.2. Nylon-6,10

Nylon-6,10 is obtained by polycondensation of equivalent amounts of hexamethylene diamine and sebacic acid chloride; the two reagents dissolved in separate phases polymerize at the interface. The chemical reaction of hexamethylene diamine and sebacic acid to form nylon-6,10 is as below;

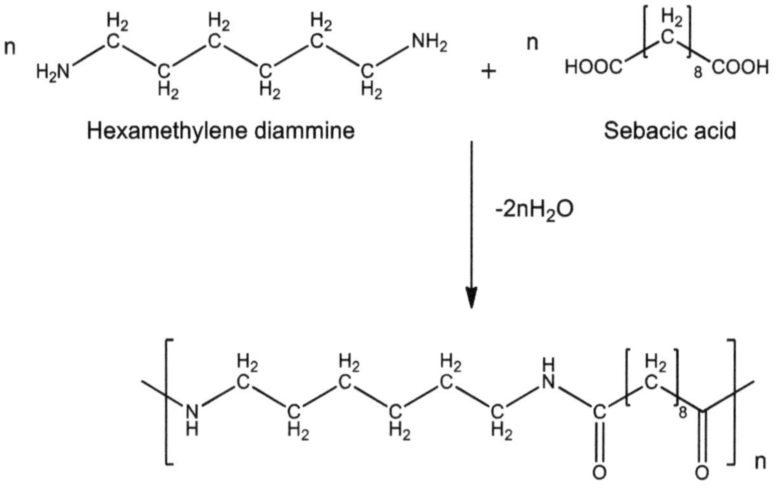

Hexamethylene diammine                    Sebacic acid

$-2nH_2O$

Nylon-6,10

## 13.2.1. Properties and Applications

Same as that of nylon-6,6.

### 13.3. Nylon-6

The preparation of nylon-6 involves ring opening-polymerization of lactams and it is a chain-growth polymerization. This polymerization process is an example of an addition polymerization with no by-products. Cyclic molecules undergo polymerization to relieve the strain; this is the main thermodynamic driving force for ring-opening during polymerization. Cyclic molecules polymerize to relieve the strain. Generally, strain in 5- and 6-membered rings is very less; therefore they do not polymerize well. But 7-membered rings, like caprolactam, are more strained and polymerize easily [24]. Nylon-6 can be prepared as follows:

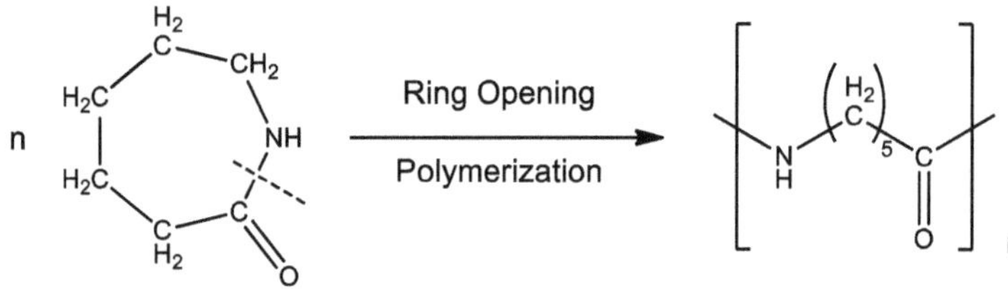

Ring Opening

Polymerization

Caprolactum

es

### *13.3.1. Properties*

Most of the properties of nylon-6, nylon-6,6, and nylon-6,10 are the same:

- Lightweight, high melting polymers
- Possess high-temperature stability
- Insoluble in common organic solvents
- Soluble in formic acid and phenol
- Flexible

### *13.3.2. Applications*

Most of the applications of nylon-6, nylon-6,6, and nylon-6,10 are same.

## 14. BAKELITE [3]

Bakelite is the first synthetic plastic developed by the Belgian chemist Leo Hendrick Baekeland in 1907. It is a thermosetting plastic also called phenol-formaldehyde and it is prepared from a condensation reaction of phenol with formaldehyde. Bakelite was designated as a National Historic Chemical Landmark on November 9, 1993, by the American Chemical Society in recognition of its significance as the world's first synthetic plastic [25].

It is prepared by condensing phenol with formaldehyde in the presence of an acidic/alkaline catalyst. This results in the formation of ortho and para hydroxymethyl phenol, which reacts to form NOVOLAC, a linear polymer.

Ortho and Para hydroxymethyl phenol

NOVOLAC

During the molding of NOVOLAC, Hexamethylenetetramine is added [$(CH_2)_6N_4$]. This converts the soluble and fusible Novolac into a hard infusible, insoluble, cross-linked polymer called Bakelite.

Bakelite

## 14.1. Properties

- Hard and rigid
- Scratch-resistant
- Resistant to water
- Resistant to chemicals except for alkalies
- Electrical insulator
- Thermal insulator

## 14.2. Applications

- Fuses and switches
- Molded articles like telephone parts, bangles, snooker balls
- Paints and varnishes
- Hydrogen exchanger resins in water softening
- Bearings

## 15. PROCESSING OF POLYMERS

Polymeric materials are formed into various three-dimensional shapes and sizes by different techniques called the processing of polymers. A variety of methods are used to process polymers and each method has its advantages and disadvantages. A specific processing technique is used to process a polymer for a specific application. Some of the important processing techniques are as follows:

1. Injection Molding
2. Blow Molding
3. Compression Molding
4. Film Insert Molding
5. Gas Assist Molding
6. Rotational Molding
7. Structural Foam Molding
8. Thermoforming

## 15.1. Injection Molding

Injection Molding is a type of polymer manufacturing process for producing parts in a very high volume with significantly varied size, complexity, and application [26]. This technique is more preferred when mass-production is required, where the same part is being created thousands or even millions of times in succession.

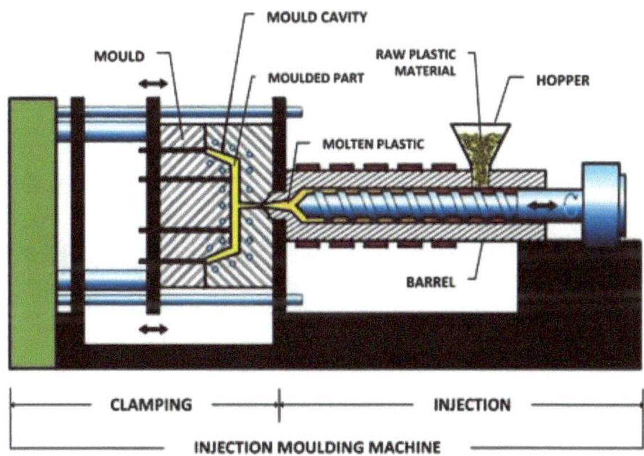


**Fig. (18).** Injection molding [26].

The injection molding process requires an injection molding machine, raw plastic/polymer material, and a mold (required shape). The plastic/polymer is melted in the injection molding machine by using heating coils and then injected into the mold tool to create the molded product where it cools and solidifies into the final part. This process generally requires a screw extruder for mixing and to melt the polymer. Then melted polymer starts accumulating in a chamber in front of the screw and once the chamber is full of molten polymer then the whole charge is ejected into the mold cavity. When the polymer in the mold is cooled to solid, then the screw can be drawn back to collect molten polymer again and this process continues. The injection molding process is considered as one of the very economical methods of processing polymers. Fig. (**18**) depicts the injection molding process [26].

## 15.2. Blow Molding

The basic principle of blow molding is inflating a hot, hollow, thermoplastic preform or parison inside a closed mold; so its shape conforms to that of the mold cavity where the material solidifies into a hollow product. The parison is a piece of tube-like plastic with a hole in one end through which compressed air can pass. The passed compressed air will push the melted polymer to match the mold structure. Once the polymer is cooled and then mold will be opened up and the part is ejected. There are mainly three types of blow molding are there;

a. Extrusion blow molding
b. Injection blow molding
c. Injection stretch blow molding.

A wide variety of hollow parts such as plastic bottles can be produced from many different plastics using this process [27]. The cost of parts produced by this molding method is not economical compared to injection molded parts but lower than rotational molded parts. Packing is the major area of application of small to medium-size disposable blow-molded products. Liquid foodstuffs are extensively packaged in narrow-neck plastic PET bottles prepared by the blow molding method. Blow-molded containers can also be used for packaging cosmetics, toiletries, pharmaceuticals, and medical goods. Fig. (**19**) shows the blow molding process [28].

**Fig. (19).** Blow molding process [28].

## 15.3. Compression Molding

Compression molding is a process in which a polymer (to be molded) is crushed into a preheated mold cavity. The mold is then closed with a force from the top as shown in Fig. (**20**) [29]. The pressure and heat are applied in such a way that polymeric material distributes uniformly through all the mold area. Heat and temperature are maintained until the polymer material is completely cured.

This method is more suitable to mold thermosetting polymers, but some thermoplastic parts can also be produced by Compression molding. Compression-molded materials are lighter, stronger, stiffer, and exhibit excellent corrosion resistance properties. Even though the production rate of compression molding is less than plastic injection molding; it produces very complex parts with more accurate geometry.

The average compression molding cycle time is about 1-6 min, which is longer than an Injection Molding cycle. This method is mainly used to fabricate large flat or reasonably curved parts used in electrical wall receptacles, brush and mirror handles, trays, circuit breakers, cookware knobs, electronic and cooking utensils, milling machine adjustment wheel, automotive parts, television cabinets, dinnerwares, radio cases, aircraft main power terminal housing, pot handles, spoilers, electric plugs and sockets [29].

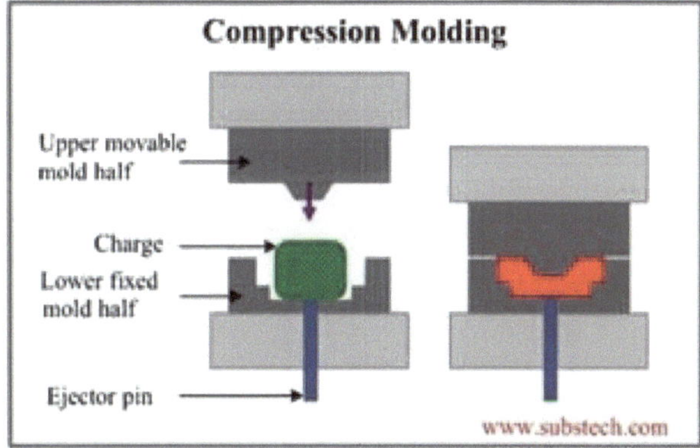


**Fig. (20).** Compression molding process [29].

## 15.4. Film Insert Molding

Film insert molding is a special type of cost-efficient injection molding; during which the high-quality parts are produced in huge numbers. This molding is extensively used to apply labels and graphics to plastic parts during the molding process. This method is generally used to fabricate polymer components very quickly with high-quality surfaces and good flexibility. By using this molding, it is possible to design the appearance of thermoplastic components, decorating the polymers, improve the surface impression.

## 15.5. Gas Assist Molding

Gas assist molding involves the injection of low pressurized nitrogen gas into the interior of a mold. Then nitrogen gas begins to flow through strategically placed gas channels to displace the material in the thick areas of the part by forming hollow sections.

The pressurized gas pushes the molten polymer resin tightly against the walls of the mold until part solidifies. The important part of this molding is maintaining the constant pressure of the gas; this results in uniform distribution of gas throughout the mold and reduces the chance of shrinking, surface blemishes, sink marks, and internal stresses of the molded parts. Fig. (**21**) depicts the working of gas-assist injection molding [30].

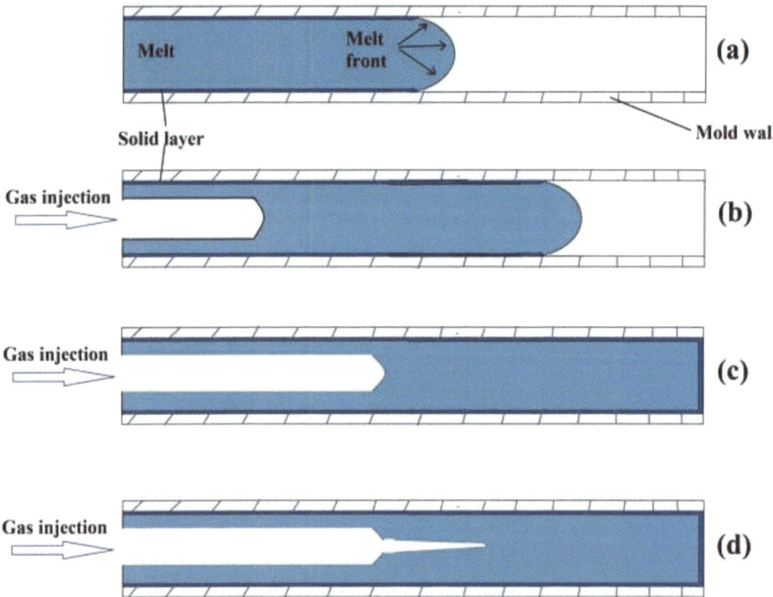


**Fig. (21).** Working of gas-assist injection molding [30].

## 15.6. Rotational Molding

Rotational molding contains a heated hollow mold filled with molten material to be mold; then it is slowly rotated to disperse the molten material until it touches the walls of the mold. The molding machines are made with a wide range of sizes and generally consist of a mold, an oven, a cooling chamber, and mold spindles. The spindles are mounted on a rotating axis, which provides a uniform coating of the plastic inside each mold [31]. Rotational molding techniques are popularly used in road construction, toy manufacturing, marine manufacturing, food and package industries, agricultural field, *etc*.

## 15.7. Structural Foam Molding

It is the process of adding a blowing agent and a slight modification of process parameters generates a sandwich construction that has a solid skin on the outer layers and a foam core in the center as shown in Fig. (**22**) [32]. This type of molding is almost similar to injection molding.

## CONCLUSION

Polymers are long-chain molecules with or without branches or crosslinks and their properties mainly depend upon chemical composition. This chapter discusses the classification of polymers based on source, structure, tacticity, properties, and monomeric units. Addition polymerization and condensation polymerization are explained in detail along with their mechanisms. The synthesis, properties, and applications of various thermoplastic and thermosetting plastics are reported in this chapter. Various processing techniques used in polymers are discussed in great detail. Knowledge of understanding the atomic or molecular structure of polymers can provide greater insight into improving them.

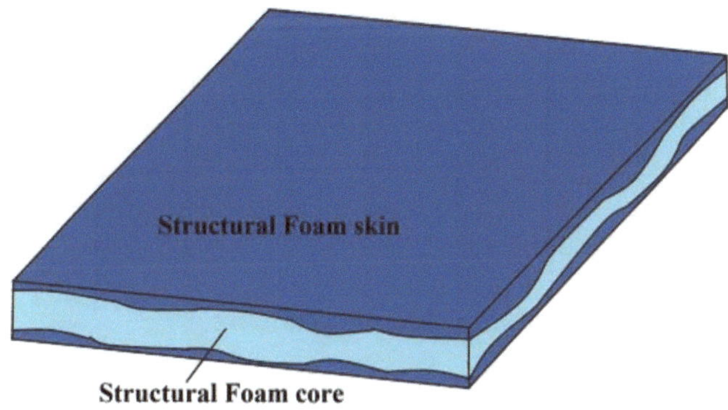


**Fig. (22).** Structural foam [32].

## QUESTIONS

1) What are polymers? Mention their properties and applications.

2) Write a note on the classification of polymers.

3) What are natural and synthetic polymers? Give some examples.

4) Explain linear, branched, cross-linked and graft polymers with examples each.

5) Define the term 'Tacticity' and mention its different types.

6) What are atactic, syndiotactic and isotactic polymers?

7) Differentiate between thermoplastics and thermosetting polymers.

8) What are homopolymers and co-polymers?

9) Define the polymerization process and explain the different types of polymerization processes.

10) Differentiate between addition and condensation polymerization.

11) Explain the various stages involved in free radical mechanism during addition polymerization.

12) Explain the various stages involved in the anionic and cationic polymerization mechanism.

13) Which is the monomer used to prepare polyethylene and mention the properties and applications of polyethylene.

14) Differentiate between low density and high-density polyethylene.

15) Explain the synthesis, properties and applications of following polymers:

(a) PVC (b) Polypropylene (c) Polystyrene (d) Nylon-6 (e) Nylon-6,6 (f) Nylon-6,10 (g) Bakelite

16) What is meant by the processing of polymers and mention different techniques used to process polymers.

17) Write a note on the following molding techniques:

a) Injection Molding

b) Blow Molding

c) Compression Molding

d) Film Insert Molding

e) Gas Assist Molding

f) Rotational Molding

g) Structural Foam Molding

## REFERENCES


[1]     *Astra Polymers.* http://www.astra-polymers.com/

[2]     A.B. Arla Plast, *Box 33, Västanavägen, SE-590 30 Borensberg, Sweden.* http://www.arlaplast.com/wp-content/uploads/2015/08/8-polymer.pdf

[3]     *Wiley Editorial, Engineering Chemistry* 2nd edWiley, .

[4]     ScienceStruck, "LDPE Vs. HDPE", https://sciencestruck.com/ldpe-vs-hdpe

[5]     Toppr, "Polymers of Commercial Importance", https://www.toppr.com/guides/chemistry/polymers/polymers-of-commercial-importance/

[6]     Wikipedia, "Graft polymer", https://en.wikipedia.org/wiki/Graft_polymer

[7]     Madhusha, "Difference Between Branched Polymer and Linear Polymer", *December 21, 2017.*

[8]     Wikipedia, *Tacticity.* https://en.wikipedia.org/wiki/Tacticity

[9]     Brandrup, Immergut, Grulke (Editors), *Polymer Handbook* 4th ed. Wiley-Interscience: New York, 1999.

[10]    Tacticity, Polymer Science Learning Center, Department of Polymer Science, The University of Southern Mississippi, https://pslc.ws/macrog/tact.htm

[11]    P.S. Stevens, *Polymer Chemistry: An Introduction.* 3rd ed. Oxford Press: New York, 1999, pp. 234-235.

[12]    A. Vashchuk, A.M. Fainleib, O. Starostenko, and D. Grande, "Application of ionic liquids in thermosetting polymers: Epoxy and cyanate ester resins", *Express Polym. Lett.,* vol. 12, pp. 898-917, 2018.
[http://dx.doi.org/10.3144/expresspolymlett.2018.77]

[13]    Engineering Chemistry, http://mgvsatya.blogspot.com/2016/09/

[14]    W.D. Callister, *Materials Science and Engineering An Introduction.* 7th ed. John Wiley & Sons, Inc, 2007.

[15]    S. Penczek, and G. Moad, "Glossary of terms related to kinetics, thermodynamics, and mechanisms of polymerization (IUPAC Recommendations 2008)", *Pure Appl. Chem.,* vol. 80, pp. 2163-2193, 2008.
[http://dx.doi.org/10.1351/pac200880102163]

[16]    Wikipedia, "Ionic polymerization", https://en.wikipedia.org/wiki/Ionic_polyme rization

[17]    "Anionic vinyl polymerization", *Polymer Science Learning Center.* https://pslc.ws/macrog/anionic.htm

[18]    "Clear Polyethylene Particles", *Product ID: CPB-0.96, Cospheric,.* https://www.cospheric.com/CPB_clear_polyethylene_polymer_spheres_2mm.htm

[19]    "Polyvinyl chloride powder", *Raghav Polymers.* https://www.exportersindia.com/raghavpolymers/polyvinyl-chloride-powder-2391174.htm

[20]    "What is Polypropylene Powder?", http://www.coloredrubber.com/blog/2012/6/4/ what-polypropylene-powder

[21]   " Uses of polypropylene, Petroquim",  http://www.petroquim.cl/usos-del-poli propileno/?lang=en

[22]   Wikipedia, "Polystyrene",  https://en.wikipedia.org/wiki/Polystyrene

[23]   Wikipedia, "Nylon 6,6",  https://en.wikipedia.org/wiki/Nylon_66

[24]   "Micro lab experiment Making Nylon6 and 6,10", *USM Polymer Science Online Laboratory Directory, Department of Polymer Science, University of Southern Mississippi*, 1998. https://www.pslc.ws/macrog/lab/lab01.htm#610

[25]   Wikipedia, "Bakelite",  https://en.wikipedia.org/wiki/Bakelite

[26]   Davies Molding, "Injection compression molding",  http://www.daviesmolding.com/Pages/Resources/Engineering-Specifications/Plastic-Molding/Injection-Compression-Molding.aspx

[27]   J.A. Schey, *Introduction to Manufacturing Processes.* 2nd ed. McGraw Hill, 1987.

[28]   Silgan Plastics, "Injection Stretch Blow Molding", http://www.silganplastics.com/capabilities-technologies/injection-stretch-blow-molding-ISBM

[29]   D. Kopeliovich, "Compression molding of polymers", *SubsTech.* http://www.substech.com/dokuwiki/doku.php?id=compression_molding_of_polymers

[30]   Sajar Plastics, "What is Gas Assist Injection Molding?",  https://sajarplastics.com/what-is-gas-assist-injection-molding/

[31]   Wikipedia, "Rotational molding",  https://en.wikipedia.org/wiki/Rotational_molding

[32]   Molded Plastics, "Structural Foam Injection Molding",  http://www.psimp.com/structural_foam_injection_molding.html

# CHAPTER 6

---

# Powder Metallurgy

**Abstract:** In this chapter, we have discussed the basic principles of Powder Metallurgy, preparation of metal powders and alloy powders by different powder metallurgy techniques and also discussed the applications of powder metallurgy. This chapter enables the student to learn how powder metallurgy has emerged as one of the superior methods to prepare metals and alloys; and the advantages and disadvantages of powder metallurgy. The use of iron metal powders was started in way back 3000 BC by Egyptians to prepare their tools for various purposes. But, the modern era of powder metallurgy started when electric bulbs were discovered, and the tungsten lamp filaments were developed by Edison. Any components can be prepared from pure metals, alloys by powder metallurgy routes. Materials like nickel, aluminum, iron, copper, brass, titanium, bronze, steels, and refractory metals are prepared by the powder metallurgy method.

**Keywords:** Atomization, Attritor mill, Blending, Ball mill, Carbonyl process, Compaction, Electrolysis process, Granulation process, Powder metallurgy, Machining, Manufacturing, Mechanical alloying, Reduction of metals, Shaker mill, Sintering, Sintering mechanism.

## 1. INTRODUCTION

Powder metallurgy is the branch of metallurgy that deals with the production of metals, non-metals, and alloy powders and the study of their structure with various applications. It also involves a series of processes like blending the fine powdered materials, their compaction and sintering to obtain the final product. Nowadays, many industries are using powder metallurgy techniques to prepare various powders [1]. It has become very popular in a very short period due to its efficiency, durability and reliable output. Fig. (**1**) shows some of the metal and alloy components prepared from powder metallurgy routes [1].

Generally, the powder metallurgy process consists of mainly four basic steps as follows [2]:

**Shashanka Rajendrachari & Orhan Uzun**

1. Manufacturing of metal powder

2. Blending of powders

3. Compacting the powders in a mold or die

4. Sintering

The schematic representation of all the above powder metallurgy steps is depicted in Fig. (**2**) [2].

**Fig. (1).** Powder Metallurgy components [1].

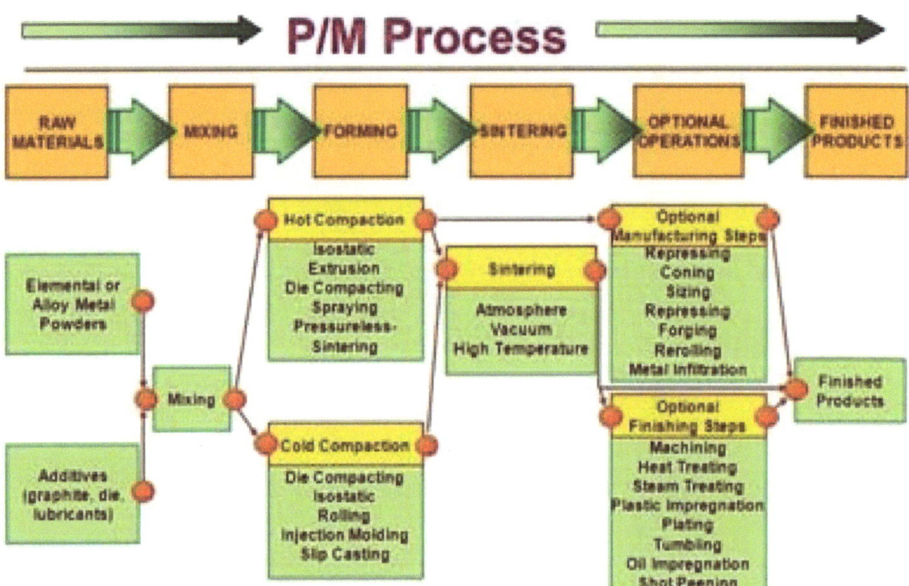


**Fig. (2).** Schematic representations of all the above powder metallurgy steps [2].

# 2. POWDER METALLURGY PROCESS

## 2.1. Manufacturing of Metal Powder

This is the first step in the processing of powder metallurgy. In this method, one can prepare various powders with different methods [3]. Each method results in different shapes and sizes of metal or non-metal powders. It is possible to control the diameter of particles from the nano range to the micron range. The shape of the powder particles mainly depends upon the type of preparation, reaction conditions, quality of raw materials, *etc*. The possible shapes of powder particles are shown in Fig. (**3**) [3].

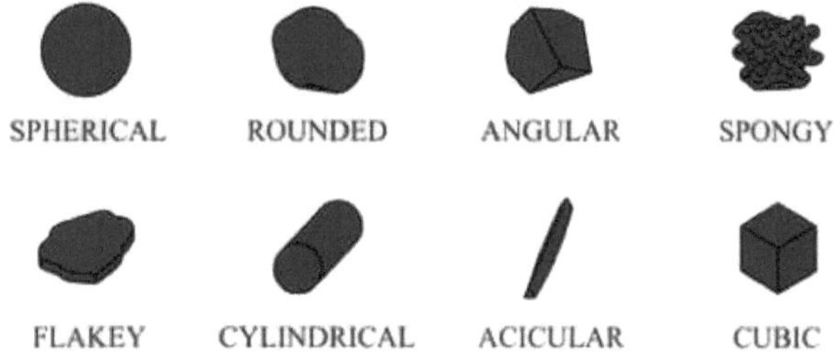


**Fig. (3).** Shapes of metal or non-metal powders [3].

The different methods used to prepare metal powders are as follows:

a. Atomization

b. Reduction of metals

c. Electrolysis Process

d. Carbonyl Process

e. Granulation Process

f. Mechanical Alloying

g. Production of fine metals by machining

### 2.1.1. Atomization

In this process, the molten metal is forced through an orifice into a stream of high-velocity air, steam or inert gas [4]. This results in rapid cooling with the disintegration of molten metal into very fine powder particles. Today, it has become one of the most suitable and well-accepted methods for producing metal powders and it is suitable for almost all metals, alloys as well as pure metals having low melting points. Metals like aluminum, iron, copper, and alloys like brass, stainless steels and superalloys can be prepared by this method. This atomization method of producing powders was used extensively during World War II. Gas atomization, water atomization, centrifugal atomization, and rotating disc atomization are important types of atomization.

### *2.1.1.1. Gas Atomization*

Gas atomization is a type of atomization process generally used to fabricate high-quality metal powders in bulk amounts. The gas atomization process is one of the less expensive methods to produce metal powders as compared to traditional techniques like crushing and grinding. This method produces irregularly shaped powder particles. The liquid metal stream is disintegrated by rapid gas expansion out of a nozzle. Commonly used gases are air, nitrogen, helium, and argon.

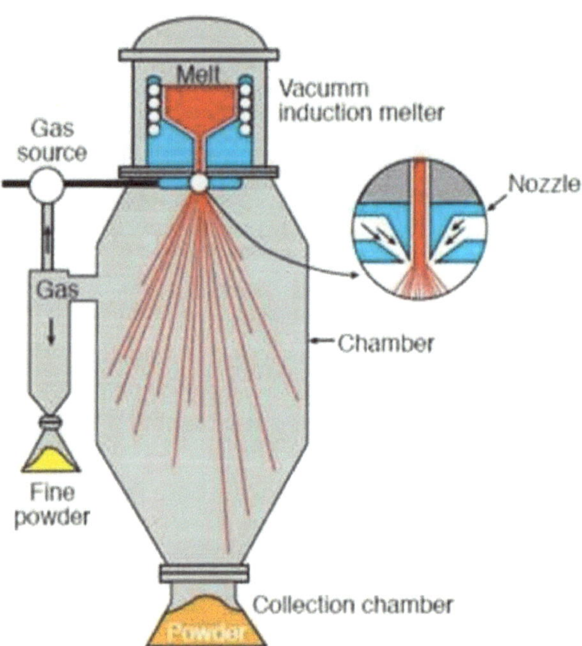


**Fig. (4).** Gas atomization tank [5].

Fig. **(4)** represents of the gas atomized process [5]. During this process, the molten metal is forced through an atomized tank and on the other hand, inert gas will be passed and the gas jets into the molten metal droplets and cools down the metal droplets to fine powder during their fall in the atomizing tank [5]. Gas atomization produces highly spherical and pure metal and alloy powders. The size and shape of the powders prepared by this method mainly depend upon the degree of heat, size of the stream, the pressure of gas and the diameter of the nozzle.

### 2.1.1.2. Water Atomization

Water atomization is almost similar to gas atomization but instead of gas, water is used as the atomizing medium. This process involves exposing falling molten metal droplets with jets of water which immediately solidify the metal into granules (>1mm) or powder (<1mm) [6]. The atomization tank in this method is slightly smaller in height as compared to the gas atomization tank. Water is generally directed by a single jet or multiple jets or annular ring around the bottom nozzle of tundish. This method involves rapid quenching; the pressure of water plays an important role in maintaining the morphology and size of the powder particles. Water has low compressibility and higher density than gas, hence the distance of the impact and the metal exit from the nozzle play less role. Fig. (**5**) shows the representation of the water atomized process [7].

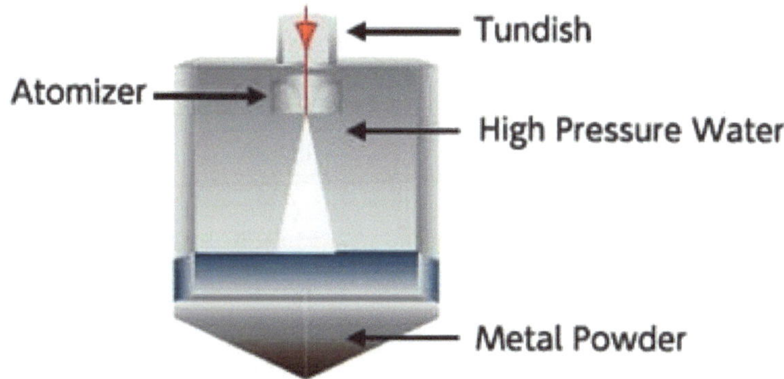


**Fig. (5).** Water atomized process [7].

### 2.1.1.3. Centrifugal Atomization

This atomization generally uses centrifugal force to break the molten metal droplets from the molten metal end of the consumable electrode. This method is more suitable to produce powders of reactive metals like titanium. In this process, one end of the metal bar is heated and melted by bringing it into contact with a non-consumable tungsten electrode, while rotating it longitudinally at high speeds. The end of a metal bar or rod is melted while it is rotated about its longitudinal axis. Molten metal produces droplets, which by centrifugal force, are ejected away from the center and solidify as spherical powder particles. The consumable rotating electrode is fed through a sealed assembly in a horizontal position located in the central axis of the tank. This electrode is an anode of the D.C. power supply [14, 15]. Fig. **(6)** depicts the diagram of centrifugal atomization [15].

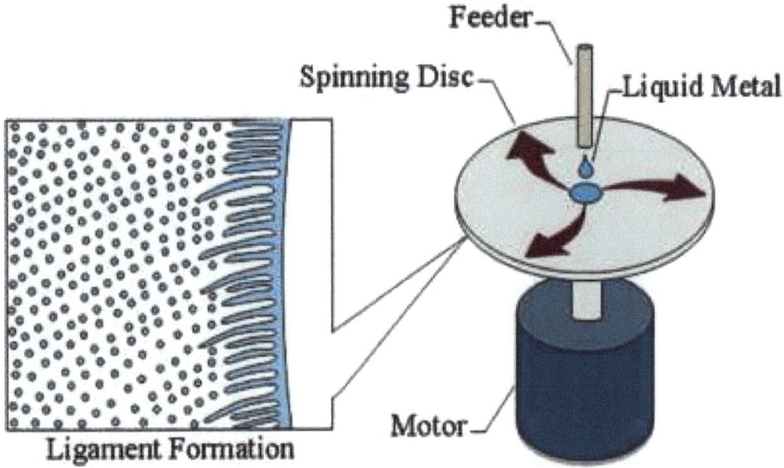


**Fig. (6).** Centrifugal atomization [15].

**Table 1. Reported range of air/liquid flow rates and present ultra-low flow rate regime.**

| Literature | Liquid Flow Rate (MG/S) | Air Flow Rate (G/S) |
|---|---|---|
| Simmons *et al.* [8] (Water) | 80-830 | 0.2-0.8 |
| Jiang *et al.* [9] (Water) | 200 | 0.4 |
| Azevedo *et al.* [10] (Water) | 120-470 | 0.037-0.238 |
| Azevedo *et al.* [11] (soy methyl ester biodiesel) | 110-560 | 0.026-0.107 |
| Azevedo *et al.* [12] (Ethanol) | 80-420 | 0.082-0.24 |
| M.A. Khan *et al.* [13] | 2.22-50.3 | 0.01-0.1 |

### *2.1.1.4. Rotating Disc Atomization*

This method involves striking a stream of molten metal onto the surface of a rapidly spinning disk. The liquid metal is mechanically atomized and thrown off the edges of the spinning disks [14]. Generally, metal powders exhibiting a low melting point can be prepared by this method. Sometimes, the disintegration of the droplets occurs after exit from the rotating disc slit. Fig. **(7)** shows the instrumental setup of rotating disc atomization [16].

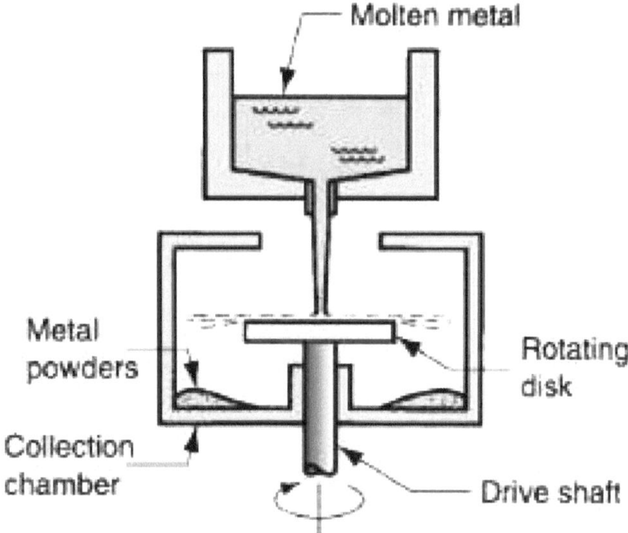


**Fig. (7).** The instrumental setup of rotating disc atomization [16].

### 2.1.2. Reduction of Metals

This process consists of grinding the metallic oxides to a fine state and subsequently, reducing it by hydrogen or carbon monoxide. This method is employed for metals such as iron, tungsten, cobalt, molybdenum, copper, *etc.* In this method, generally, the metal oxide is reduced to its metal powder in the presence of a reducing gas below its melting temperature. For example, if iron ore is crushed and mixed with carbon and passed through a continuous furnace, a reduction reaction takes place, leaving a cake of sponge iron. Then, sponge iron is crushed and sieved to produce iron powder. The purity of the powder mainly depends upon the raw materials.

### 2.1.3. Electrolysis Process

Electrolysis is a method of producing powders either in an electrolyte or as a deposition [17]. This method mainly depends upon the composition, temperature, and strength of the electrolyte. By using this method, we can deposit many metals either in a spongy or in a powdery state. Sometimes, secondary operations like washing, drying, reduction, annealing, and crushing may be required. Copper is the most common metal that can be produced by this method, but nowadays, people are producing chromium, manganese as well by electrolysis. This method is expensive but produces high purity and high-density powders with significant powder morphology. Fig. (**8**) shows the electrolysis instrumental setup to prepare copper powder deposition [17].

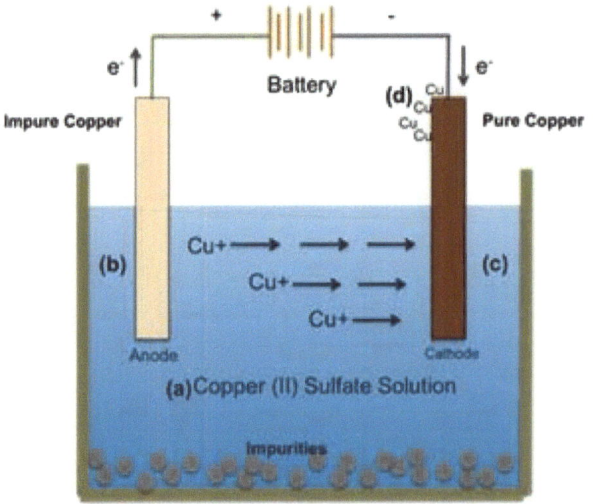


**Fig. (8).** Experimental setup of electrolysis to prepare copper powder deposition [17].

### 2.1.4. Carbonyl Process

In this method, metals are allowed to react with carbon monoxide to form the respective metal carbonyls. Then metal carbonyls are decomposed at a higher temperature to produce high purity metals. Metals like iron and nickel are produced in bulk amount by decomposition of the metal carbonyls. This method generally results in small, uniform spherical particles typically 5 microns in diameter [18]. This method produces high purity powders but it is an expensive method. This method is significantly used to refine the impure nickel. Initially, nickel is allowed to react with carbon monoxide to form nickel carbonyl gas $(Ni(CO)_4)$. Then at a suitable high-temperature, nickel carbonyl gas can be decomposed to obtain pure nickel metal and carbon monoxide will be recovered [19]. The purity of nickel can be increased further by using thermal shock decomposition. Refineries in North America and Britain are producing around 50,000 tonnes of nickel per year using this method, as this method is economical [19]. Another major benefit of this method involves no real waste products, as used gas can also be recycled back into the main refinery process.

### 2.1.5. Granulation Process

Granulation is an important area for powders within the ceramic and the powder metallurgy area which enhances the flow properties of powders. Granulation process is the process of collecting particles together by creating bonds between them. Granules are formed from the powder particles by wetting and nucleation, coalescence or growth, consolidation, and attrition or breakage [20]. The granulation process involves pressing one or more powder particles within a required limit to form a granule. This process can significantly improve the overall quality of pressed parts and also increases the density of the powder prior pressing. Resulted in high-density powders prior compaction enhances the density significantly after compaction due to the formation of stronger bonds after pressing. Granulation technique is classified into two types:

I. Dry granulation: Involves formation of granules without using a liquid solution

II. Wet granulation: Involves formation of granules with using a liquid solution

Wet granulation is one of the most popular and widely used granulation techniques [20].

### 2.1.6. Mechanical Alloying

Mechanical alloying is a solid-state and powder processing technique that involves flattening, cold welding, fracturing, and re-welding of powder particles in a high-energy ball mill to produce a homogeneous solid solution materials. The transfer of mechanical energy to the powder particles results in the introduction of strain into the powder through the generation of dislocations and other defects which act as fast diffusion paths [21 - 27]. Additionally, refinement of particle and grain sizes occurs, and consequently, the diffusion distances are reduced. Mechanical alloying technique was first developed in the mid-1960 by John Benjamin to produce nickel-based oxide dispersion strengthened superalloys for gas turbine applications [28].

Fig. (**9**) represents the mechanical alloying process [29]. Different types of high-energy milling instruments are used to produce mechanically alloyed or milled powder. They are different in size, shape, efficiency as compared to one another and they are explained as follows:

### 2.1.6.1. SPEX Shaker Mills

These are high energy ball mills generally used to grind, pulverize or mix powder samples to the required limit. SPEX shaker mills are most commonly used to produce only 10-20g of powder at a time and hence are mainly used for laboratory investigation and alloy screening purposes. This mill has one vial, containing the powder sample and grinding balls, secured in the clamp and swing energetically back and forth several thousand times a minute [29]. The back-and-forth shaking motion is combined with lateral movements of the ends of the vial so that the vial moves like a number 8 or infinity symbol. During the vial movement, the balls inside the vials will strike against the sample and the sidewalls of the vial with great impact force. Fig. (**10**) shows the SPEX shaker mill [30]; it is generally used for both milling and mixing the sample. These mills are considered high energy ball mills due to the high amplitude (about 50 mm) and high speed (about 1200 rpm) [30].

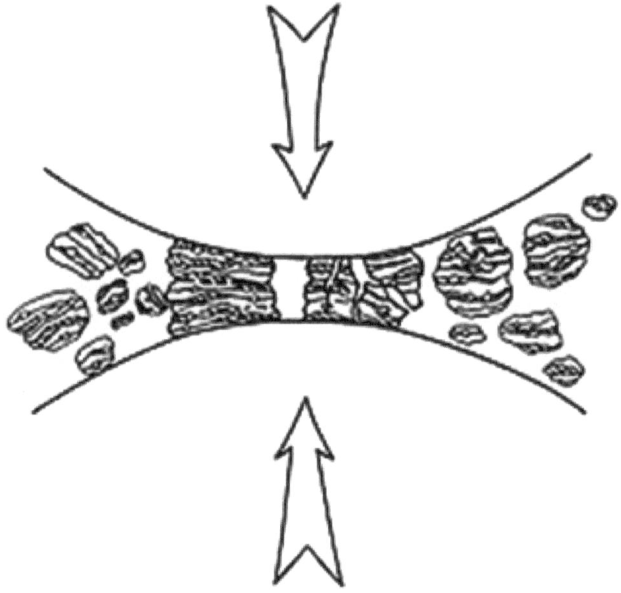

**Fig. (9).** Mechanical alloying [29].

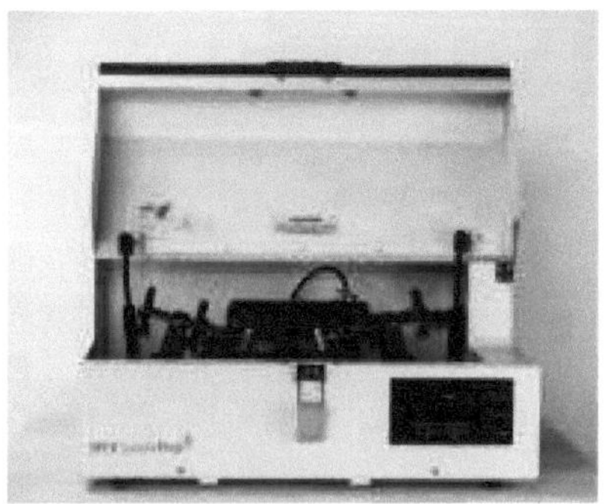

**Fig. (10).** SPEX shaker mill [30].

## 2.1.6.2. Planetary Ball Mills

Planetary Ball Mills are used to attain the highest degree of fineness. Due to the high centrifugal forces of a planetary ball mill result in very high pulverization energy and therefore short grinding times. It is a type of grinder used to grind and blend materials especially used in mineral dressing processes, paints, pyrotechnics, ceramics, *etc.* A few hundred grams of the powder can be milled at the same time.

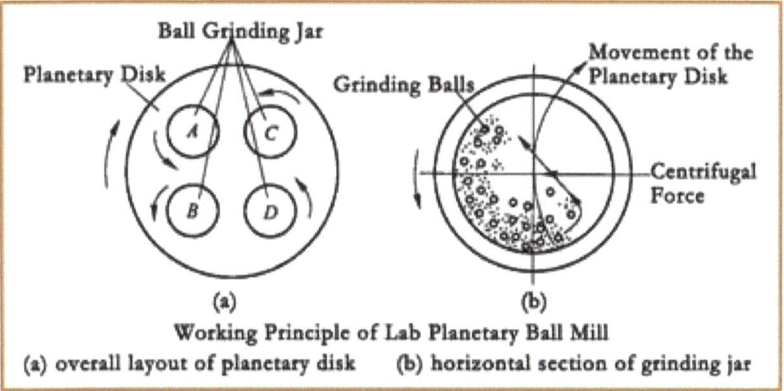


**Fig. (11).** The working principle of planetary ball mill [31].

The name planetary ball mill is due to the planet-like movement of its vials; these vials are arranged on a rotating disk and a special drive mechanism causes them to rotate around their axes [31]. This movement generates centrifugal force and refines the powders. Both the vials and the supporting disk rotate in the opposite direction, the centrifugal forces alternately act in like and opposite direction [32]. Planetary ball mills not only produce centrifugal force but also acceleration forces and Coriolis forces and these lead to a strong grinding effect between the grinding balls and the sample. Fig. (**11**) shows the working principle of a planetary ball mill [31].

## 2.1.6.3. Attritor Mills

This ball mill consists of a rotating horizontal drum that contains almost 50% filled small steel balls and a series of impellers. The impellers energize the charged balls and powder particles; and they undergo a series of various phenomenons such as impact, rotation, tumbling, and shear. This results in the reduction of the particle size due to the collisions between balls-container wall-agitator-shaft-impellers. Therefore, particle size reduces to the micro-level and sometimes to the nano level.

As the drum starts rotating, then the balls begin to drop on the metal powder with greater impact and efficiently ground the powders. The rate of grinding increases with the speed of the drum rotation [33]. At higher speed, the centrifugal force acting on the steel balls exceeds the force of gravity and the balls are pinned to the wall of the drum and no grinding will take place-this action is called pinning. An attritor (a ball mill capable of generating higher energies) consists of a vertical drum containing a series of impellers. By using this mill large quantities of the powders can be produced. The velocity of the grinding medium in attritor mill is much lower than in the planetary or SPEX mills. Therefore, attritor mills are considered as low energy ball mills. Fig. (**12**) depicts the attritor mill [33].

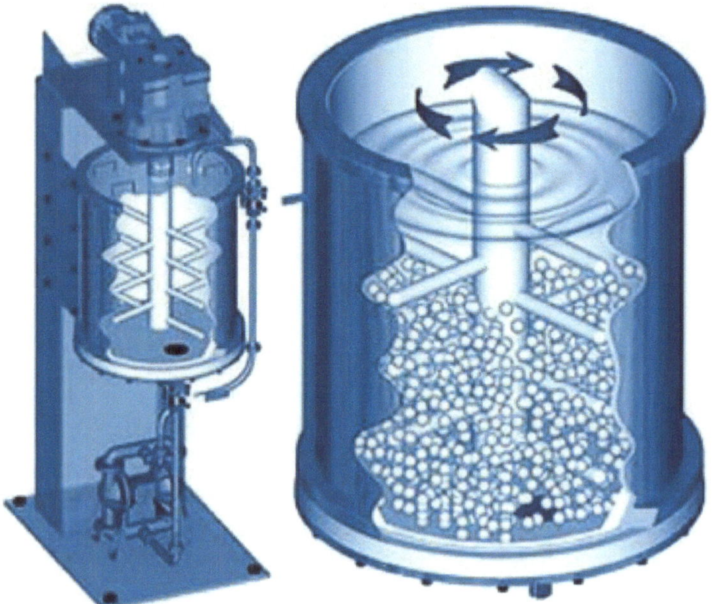

**Fig. (12).** Attritor mill [33].

### *2.1.7. Production of Fine Metals by Machining*

A machining process of fine metals production can be carried out by using a lathe or a milling cutter in which something more than just scratching is involved. In this method, the attacking tool digs under the surface of the metal and tears it off.

It is possible to produce fine chips in bulk amounts by applying small force on a lathe machine [34]. The produced scrap from lathe machine is in the form of chips and turnings and their size can be further refined by using grinding or ball mills. Fig. (**13**) shows a lathe machine. Machining is a useful process when consuming the scrap from another process [34, 35]. But, it is difficult to control the powder characteristics due to the possibility of chemical contamination such as oxidation, oil, and other metal impurities [34, 35]. The particles produced by this method exhibit irregular shape and coarse nature.

**Fig. (13).** Lathe machine [34, 35].

## 3. BLENDING OF POWDERS

In many industrial processes blending of powder particles is a common process and performed in large quantities more frequently. In general, blending is an act of combining or mixing materials to form homogeneous mixtures; however, this operation usually occurs gently with multiple components. Mixing flour, salt, yeast, and water to make bread is a very simple example of blending. Pharmaceutical and Food Industries more depend upon blending technology as small amounts of a powdered active drug are carefully blended with excipients such as starch, cellulose, lactose, and lubricants. In the food industry, many powder products result from blending technology are cake mix, ice tea, flavored drinks and curry [36].

Blending a coarser fraction with a finer fraction ensures that the interstices between large particles will be filled out and below are some of the characteristics of blending.

- Powders of different metals and other materials may be mixed to impart special physical and mechanical properties through metallic alloying.

- To improve the flow of powders, lubricants may be mixed to improve the powder's flow characteristics.

- Binders such as wax or thermoplastic polymers are added to improve green strength.

Convection, diffusion, and shear are the three primary mechanisms of blending. Convection blending depicts the gross movement of particles through the mixer either by a force action from a paddle or by gentle tumbling under rotational effects. Similarly, diffusion is a slow blending mechanism. Lastly, the shear mechanism of blending involves a thorough incorporation of material passing along forced slip planes in a mixer [36].

Some of the blenders are more suitable for dry particles while few of them are best suited for gentle blending. Fig. (14) shows the different types of blenders used in powder metallurgy [36]. The following is a list of different types of blenders used in chemical processing plants.

a. Ribbon Blender
b. Cone Blender
c. Hydraulic Blender
d. Paddle Blender
e. Vertical Blender
f. Tumble Blender

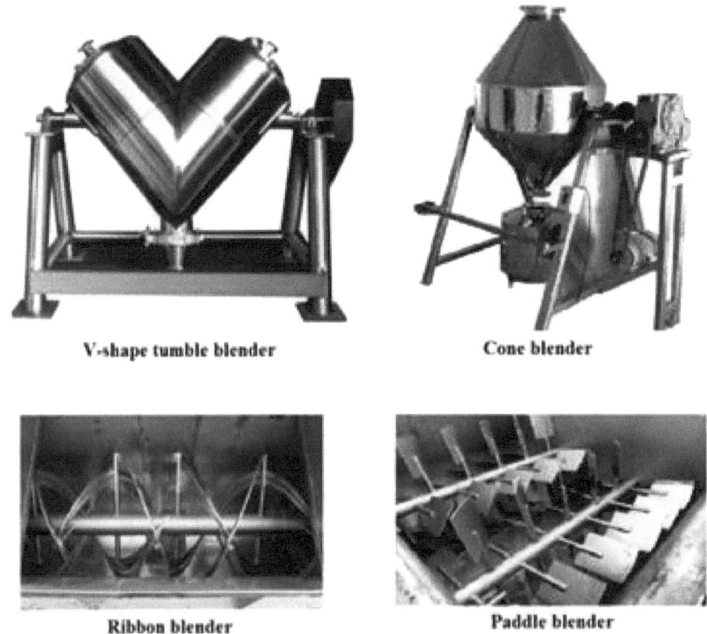


**Fig. (14).** Different types of blenders used in powder metallurgy [36].

## 4. COMPACTING THE POWDERS IN A MOLD OR DIE

Powder compaction is the process of pressing a specific amount of metal powder in a die by applying high load (100-1000MPa); then powder in the die is compacted into a required shape and then ejected from the die cavity. The load required for compaction mainly depends upon the characteristics and shape of the particles, shape, and size of the die used, blending or mixing method used before compaction and on the lubricant used. Compaction is generally performed at room temperature; during compaction, the loose powder is consolidated and densified into a specifically shaped compact called as green compact [37]. As the green compact comes out of the die, the compact will have the shape and size of the finished product. The green compact exhibits low strength but it is sufficient for in-processes safe handling and transportation to the sintering furnace. Fig. (15) depicts the typical set of powder metallurgy compaction tools [37].

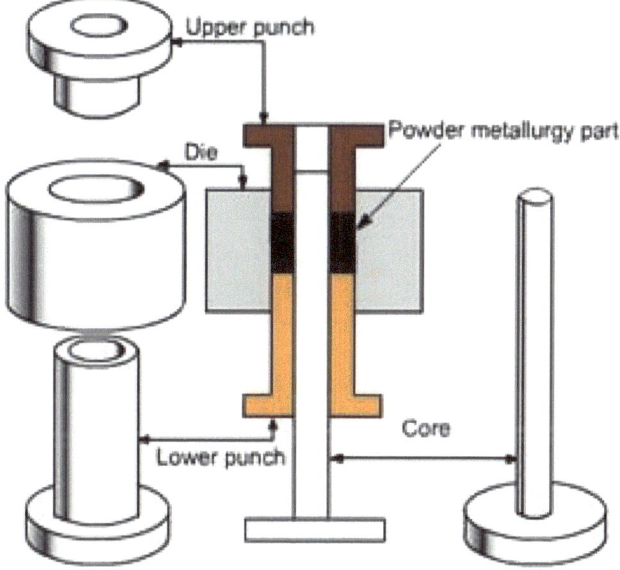


**Fig. (15).**  Typical set of powder metallurgy compaction tools [37].

Compaction or pressing of metal powders is mainly divided into the following two types:

a. Hot Pressing

b. Cold Pressing

Hot pressing is one of the widely used methods of compaction of powder at higher temperatures, and it involves the simultaneous application of both load and sintering. The hot pressing technique is used mostly in the manufacturing of carbide cutting tools and in a few specialized applications [38].

Similarly, cold pressing can be also be further classified into the following processes:

- Axial pressing or conventional pressing.

- Isostatic pressing: It is a technique where pressure is applied uniformly to the metal powders.

- Cold pressing: It is the method of applying pressure upon a column of loosely bound metal powders in a closed die to form a green compact.


## 5. SINTERING

Sintering is a heat treatment performed for a powder compact (green compact) to impart strength and integrity [39]. The sintering temperature is always should be well below the melting point of the major constituent of the Powder Metallurgy material. After compaction, the powder particles are held together by cold welds, therefore, the green compact exhibits low strength but which give the compact sufficient "green strength" to be handled. During sintering, the material transformation will take place between two neighboring particles through diffusion phenomenon and this result in neck formation at these contact points.


### 5.1. Sintering in The Powder Metallurgy Process

There are mainly three stages of solid-state sintering:

  i. Initial stage,

 ii. Intermediate stage,

iii. Final stage are shown in the Fig. (**16**) [39].

Generally, solid-state sintering takes place through following two mechanisms [39]:

i. Removal of the lubricant (added during compaction) by evaporation and burning of the vapors.

ii. Reduction of the surface oxides from the powder particles in the compact.

These mechanisms take place during the sintering process by using an appropriate temperature profile throughout the furnace.

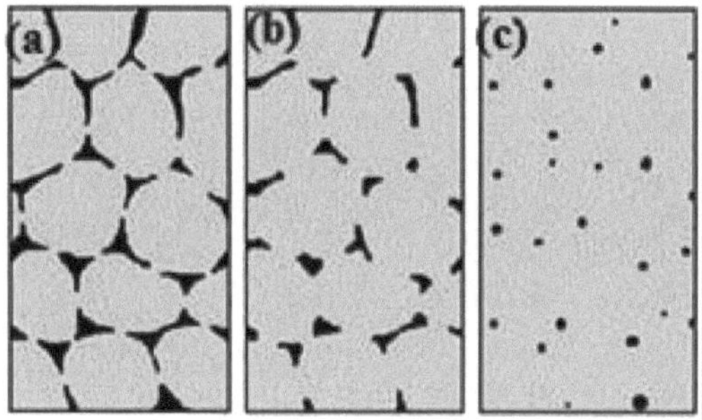

**Fig. (16).** The three stages of solid-state sintering **(a)** Initial stage, **(b)** Intermediate stage, **(c)** Final stage [39].

## 5.2. Sinter Hardening

During the sintering process, cooling plays an important role in maintaining the microstructure. Now a day sintering furnaces with accelerated cooling rates are available and material grades have been developed that can transform to martensitic microstructures at high cooling rates. This process, together with a subsequent tempering treatment, is known as sintering hardening. This process is developed recently and leads to enhanced sintered strength.

## 5.3. Liquid Phase Sintering

### 5.3.1. Transient Liquid Phase Sintering

During solid state sintering process compact will undergo some degree of shrinkage as the sintering necks grow. However, a common practice with ferrous PM materials is to make an addition of fine powders whose melting temperature is significantly less (copper or aluminum) to create a transient liquid phase during sintering. Fig. (**17**) represents the solid phase and liquid phase sintering [40].

At sintering temperature, the fine powder with low melting temperature melts and then diffuses into the iron powder particles creating swelling [40]. By careful selection of copper content, it is possible to balance this swelling against the natural shrinkage of the iron powder skeleton and provide a material that does not change in dimensions at all during sintering. The aluminum addition also provides a useful solid solution strengthening effect [39].

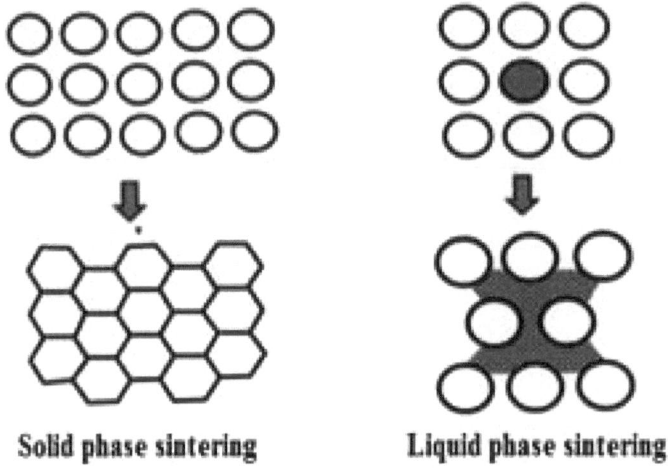


**Fig. (17).** Comparison of solid phase and liquid phase sintering [40].

### 5.3.2. Permanent Liquid Phase Sintering [41]

Some special materials like cemented carbides or hard metals involve a different sintering mechanism slightly differ from the normal liquid sintering method. This method involves the use of an additive to the powder and this generates a permanent liquid phase. Generally, the added additive will melt before the matrix phase and which will often create a so-called binder phase. The process has three stages and shown in Fig. (**18**) [41].

#### 5.3.2.1. Rearrangement

When liquid melts, pores present in the compact will pull the liquid through capillary action and cause grains to rearrange into a more favorable packing arrangement.

#### 5.3.2.2. Solution-precipitation

In areas where capillary pressures are high, atoms will preferentially go into solution and then precipitate in areas of lower chemical potential where particles are not close or in contact. This is called contact flattening and densifies the system in a way similar to grain boundary diffusion in solid-state sintering [39].

### *5.3.2.3. Final Densification*

Densification of the solid skeletal network, liquid movement from efficiently packed regions into pores.

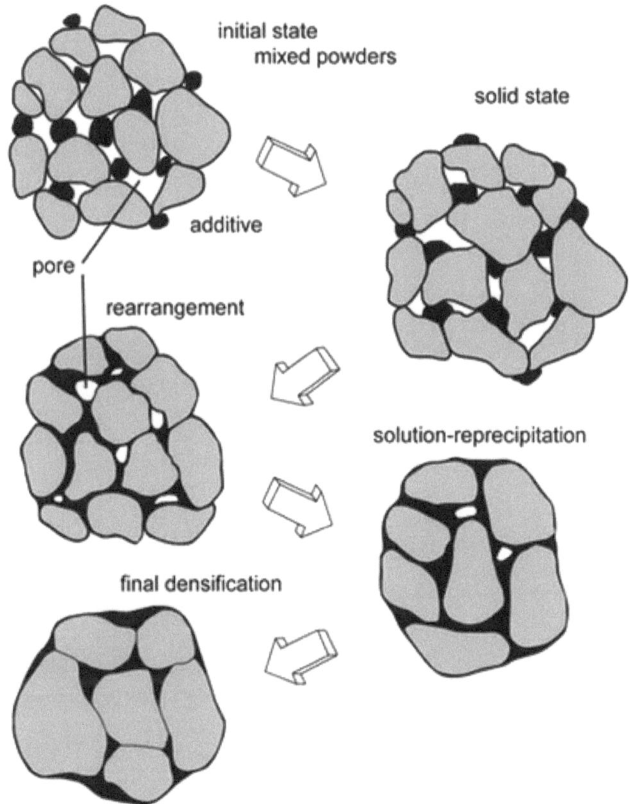


**Fig. (18).** The change of microstructure during liquid phase sintering [41].

## 6. APPLICATIONS OF POWDER METALLURGY

### 6.1. The Automotive Sector

The automotive sector is having a huge market for sintered structural powder metallurgy parts; around 80% of all PM structural components are used alone for automotive applications. Automotive applications include transmissions (automatic and manual) and engines. Transmission components like synchronizer system parts, gear shift components, clutch and turbine hubs, planetary gear carriers, clutch, and pocket plates. Engine parts include pulleys, sprockets and hubs, valve seat inserts, valve guides, assembled camshafts, balancer gears, engine manifold actuators, engine management sensor rings, *etc*. Fig. (**19**) shows the automotive applications of PM parts [42]. PM parts can also be used in oil pumps – particularly gears, shock absorbers – piston rod guides, piston valves, end valves, anti-lock braking systems (ABS) – sensor rings, chassis components, exhaust systems – flanges, oxygen sensor bosses, turbochargers [42].

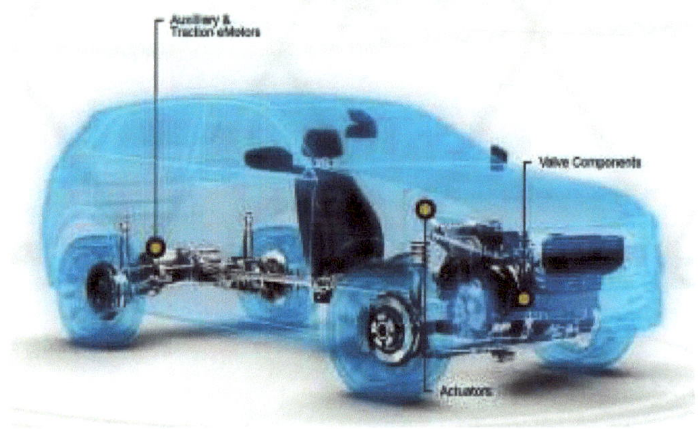


**Fig. (19).** Automotive applications of PM parts [42].

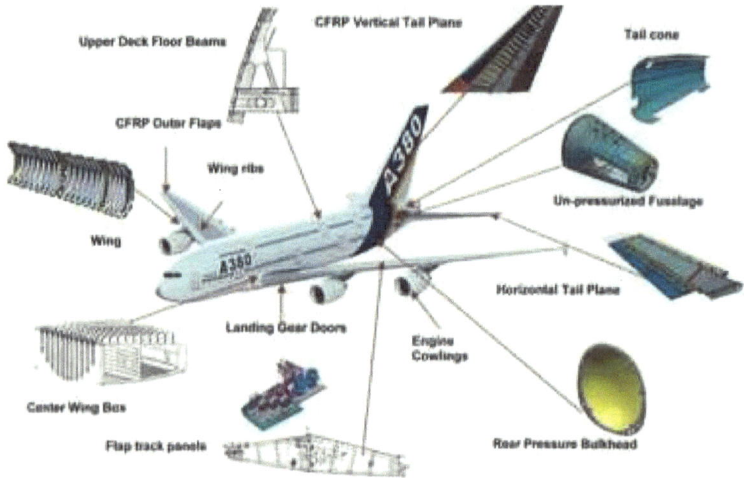


**Fig. (20).** Aerospace applications of PM parts [43].

## 6.2. Aerospace Applications

PM parts can be used in aero-engine as shown in the Fig. (**20**) [43] and land-based gas turbine applications. Parts used in aerospace applications require extremely good properties; powder metallurgy parts exhibit all the good properties due to the use of a new sintering technique called as Hot Isostatic Pressing (HIP). This HIP results in high strength PM parts.

## 6.3. Oil and Gas Industry

Hard metal and diamond cutting tools are regularly used for oil and gas exploration. They are also used for machining operations in many industrial sectors including automotive, aerospace and general engineering.

## 6.4. Healthcare Sector

The healthcare sector has many devices that contain PM components. MRI scanners contain large quantities of permanent magnets processed from powders. Most of the surgical instruments and dental implants are produced by Metal Injection Moulding. PM materials are used in porous implant structures to match bone stiffness and to aid osteointegration.

## 6.5. Filters

PM filters possess greater strength, porosity and shock resistance than ceramic filters.

## 6.6. Cutting Tools and Dies

Cemented carbide powder can be used as cutting tools; graphite powder is used in the fabrication of dies.

## 6.7. Magnets

Magnets are produced from different compositions of powders of iron, aluminum, nickel, and cobalt.

## 6.8. Turbine Disk Super Alloy

The turbine disk is the primal rotating part of aero-engine which exhibit maximum load-bearing capacity for a long time at high temperature and pressure. Therefore, the materials used of turbine disk should possess excellent mechanical properties at high temperatures. These turbine disks are made up of nickel-based alloy. Recently powder metallurgy method has proved to be the best metallurgical method to prepare turbine disc with excellent mechanical properties, precise design, high temperature resistance, *etc.* Researchers are trying for the further improvement of the PM turbine disc by electroplating with cubic boron nitride and use of super-hard abrasive grinding technology. Hence, it is of great importance to study the point grinding of cubic boron nitride electroplating on PM turbine disc for improving the surface quality of fine parts of the turbine disk (see Fig. **21**) [44] and promoting the manufacture of new aero-engines.

**Fig. (21).** Structure of PM turbine disk [44].

## 6.9. Injection Molded 3C Products

Most of the plastic components available in the market are injection molded. Currently some of the metals and alloy powders are also injection molded for automobile applications. The powder injection molding process involves blending or mixing of the powdered metals followed by pelleting, injection, de-binding, sintering, and post processes. The injection molded products are used in variety of fields, such as electronics, automotive parts, medical parts, precise parts, *etc*. The advantages of using powder injection molding is high design flexibility, abundant raw material choices, excellent physical and chemical properties, delicate cosmetic, high dimension accuracy, strong and flexible mass production capability, environment friendly manufacturing [45].

## 6.10. Sintered Soft Magnetic Materials

Soft magnetic materials are mainly used in electromagnetic devices like electric motors, sensors, actuators, generators. The soft magnetic materials produced by powder metallurgical technology exhibits excellent three-dimensional (3D) isotropic magnetic behavior, production flexibility of material's design and environment friendly [46]. There are mainly two kinds of soft magnetic materials; sintered soft magnetic materials and soft magnetic composites. Sintered soft magnetic materials are produced by uniaxial pressing and sintering of alloyed powder (generally, pure iron, iron phosphorus, iron nickel, iron silicon, iron cobalt and ferritic stainless steels are used) at elevated temperature. Whereas, soft magnetic composites are made up of insulated iron powder [46, 47]. Fig. (**22**) depicts the sintered soft magnetic materials [47].

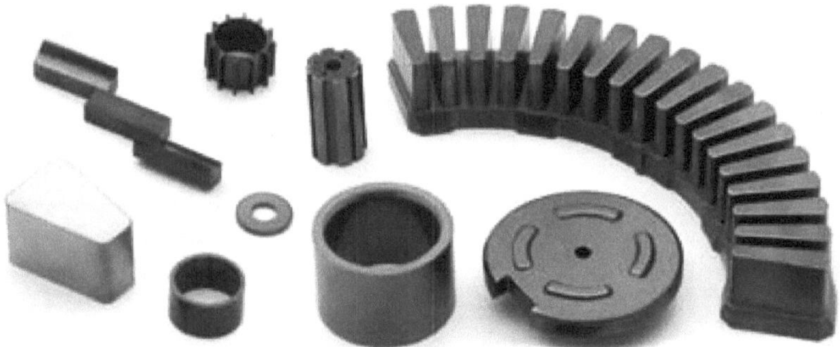

**Fig. (22).** Sintered soft magnetic materials [47].

# 7.  ADVANTAGES  AND  DISADVANTAGES  OF  POWDER METALLURGY [48]

## 7.1. Advantages

1. Powder metallurgy produces near-net-shape components.

2. Most of the powder metallurgy technique does not require secondary operations or may require few.

3. It is possible to produce powder metallurgy parts from high melting point refractory metals with ease and less cost.

4. Powder metallurgy does not require further machining.

5. It involves a high production rate with a low unit cost.

6. It can produce even complicated structures with a uniform microstructure.

7. Different combinations of materials can be used in PM products, which are otherwise impossible to make. For example, mixing ceramics with metals (Cermets).

8. It has flexibility for producing PM parts with specific physical and mechanical properties like hardness, strength, density, and porosity. In other words, we can control all these properties just varying the composition.

9. By using powder metallurgy, parts can be produced with infiltration and impregnation of other materials to obtain special characteristics that are needed for specific and special applications.

10. Powder metallurgy can be used to produce bi-metallic products, porous bearing, and sintered carbide.

11. Powder metallurgy makes use of 100% raw material as no material is wasted as scrap during process.

12. It provides properties like porosity and self-lubrication to the manufactured parts.

## 7.2. Disadvantages

1. The PM-products have limited shapes and features.

2. This technique causes potential workforce health problems from atmospheric contamination of the workplace due to the particulate pollution.

3. The types of equipment required for powder metallurgy are very expensive, which means the initial set up of tools and equipment is very much expensive.

4. It is difficult to produce large and complex shaped parts with powder metallurgy.

5. The PM parts exhibit low ductility and strength.

6. Most of the finally divided powders like aluminum, magnesium, titanium, and zirconium are fire hazard and explosive. Decreased particle size increases the reactiveness of the powder particles.

7. This technique is not useful for low melting powder such as zinc, cadmium, and tin as they show thermal difficulties during sintering operations.

8. The cost of raw material is very high.

9. Mechanical properties of the PM parts are relatively low quality as compared to cast or machined parts.

10. Sometimes, the density of different parts of the final product can vary due to uneven compression.

## CONCLUSION

The powder metallurgy involves fabrication of materials which exhibits greater strength, hardness, and uniform microstructures. It also involves low or no wastage, no secondary operations, near net shape components, different combination of components, *etc*. This chapter describes various methods of preparing metal or alloy powders. Blending, compaction and sintering are explained in detail. Different types of sintering and applications of powder metallurgy are discussed in the present chapter. We also discussed the advantages and dis-advantages of powder metallurgy in detail.

## QUESTIONS

1) What is powder metallurgy?

2) Explain the basic steps in Powder Metallurgy.

3) Mention some of the possible shapes of powders produced from powder metallurgy.

4) Mention the different methods used to prepare metal powders.

5) What is atomization? Explain various types of atomization.

6) Explain the electrolysis method of powder production.

7) What is granulation? Explain different types of granulation?

8) What is mechanical alloying? Explain different types of ball mills.

9) Why the planetary ball mills are named so? Explain the principle of planetary ball mills.

10) Define blending. Write some of the characteristics of blending.

11) Mention different mechanisms and various types of blending.

12) What is compaction?

13) What is a green compact?

14) Write a note about hot pressing and cold pressing.

15) What is sintering? Explain the two mechanisms involved in solid-state sintering.

16) What is sinter hardening?

17) What is the liquid phase sintering? Explain the various stages present in the liquid phase sintering.

18) Write some of the applications of powder metallurgy.

19) What are the advantages and disadvantages of powder metallurgy?

## REFERENCES

[1]     TPI Powder Metallurgy, https://www.tpipm.com/

[2]     A. Carr, https://slideplayer.com/slide/10352339/

[3]     Inside Metal Additive Manufacturing, "The role of (super) powders in SLM", 2014.

[4]     Module 3, "Selection of Manufacturing Processes", *Lecture 6, Design for Powder Metallurgy.* https://nptel.ac.in/courses/112101005/downloads/Module_3_Lecture_6_final.pdf

[5]     D.P. Guillen, J.P. Wharry, and D.W. Gandy, "Neutron Irradiation of Nuclear Structural Materials Fabricated by Powder Metallurgy with Hot Isostatic Pressing", *American Nuclear Society Annual Meeting,* 2017 San Francisco, CA

[6]     Total Materia, "The Water-Atomizing Process: Part One", *May,* 2013. https://www.totalmateria.com/page.aspx?ID=CheckArticle&site=ktn&NM=309

[7]     Epson Atmix Corporation, "Atomization Technology",

[8]     B.M. Simmons, H.V. Panchasara, and A.K. Agrawal, "A comparison of air-blast and flow-blurring injectors using phase doppler particle analyzer technique", Proceedings of the ASME Turbo Expo, 2009.
[http://dx.doi.org/10.1115/GT2009-60239]

[9]     L. Jiang, and A.K. Agrawal, "Spray features in the near field of a flow-blurring injector investigated by high-speed visualization and time-resolved PIV", *Exp. Fluids,* vol. 56, pp. 1-13, 2015.
[http://dx.doi.org/10.1007/s00348-015-1973-z]

[10]    C.G. Azevedo, J.C.D. Andrade, and F.D.S. Costa, "Effects of nozzle exit geometry on spray characteristics of a blurry injector", *At. Sprays,* vol. 23, pp. 193-209, 2013.
[http://dx.doi.org/10.1615/AtomizSpr.2013007244]

[11]    C.G. Azevedo, J.C.D. Andrade, and F.D.S. Costa, "Effects of injector tip design on the spray characteristics of soy methyl ester biodiesel in a blurry injector", *Renew. Energy,* vol. 85, pp. 287-294, 2016.
[http://dx.doi.org/10.1016/j.renene.2015.06.021]

[12]    C.G. Azevedo, F.D.S. Costa, and J.C.D. Andrade, "Experimental valuation diagnostics of hydrous ethanol sprays formed by a blurry injector", *J. Aerosp. Technol. Manag.,* vol. 5, pp. 197-204, 2013.
[http://dx.doi.org/10.5028/jatm.v5i2.231]

[13]    M.A. Khan, H. Gadgil, and S. Kumar, "Influence of liquid properties on atomization characteristics of flowblurring injector at ultra-low flow rates", *Energy,* vol. 171, pp. 1-13, 2019.
[http://dx.doi.org/10.1016/j.energy.2019.01.006]

[14]    A. Bond, "Physical Methods: Suitable for comparatively less reactive metals", https://slideplayer.com/slide/5801459/

[15]    P. Sungkhaphaitoon, S. Wisutmethangoon, and T. Plookphol, "Influence of Process Parameters on Zinc Powder Produced by Centrifugal Atomisation", *Mater. Res.,* vol. 20, no. 3, pp. 718-724, 2017.
[http://dx.doi.org/10.1590/1980-5373-mr-2015-0674]

[16]    Encyclopedia    of    Engineering,    "Powder    Metallurgy,    Atomization",
        http://www.mechscience.com/atomization-powder-metallurgy/

[17]    Study.com, Chapter 3, Lesson 16, "What is Electrolysis? - Definition, Process & Facts",
        https://study.com/academy/lesson/what-is-electrolysis-definition-process-facts.html

[18]    G. Greetham, AZoM, "Powder Metallurgy - Powder Production",

[19]    J. Jones, AZoM, "Nickel Powders from the Carbonyl Process",

[20]    S. Shanmugam, "Granulation techniques and technologies: recent progresses", *Bioimpacts,* vol. 5, no.
        1, pp. 55-63, 2015.
        [http://dx.doi.org/10.15171/bi.2015.04] [PMID: 25901297]

[21]    R. Shashanka, and D. Chaira, "Phase transformation, and microstructure study of nano-structured
        austenitic and ferritic stainless steel powders prepared by planetary milling", *Powder Technol.,* vol.
        259, pp. 125-136, 2014.
        [http://dx.doi.org/10.1016/j.powtec.2014.03.061]

[22]    R. Shashanka, and D. Chaira, "Development of nano-structured duplex and ferritic stainless steel by
        pulverisette planetary milling followed by pressureless sintering", *Mater. Charact.,* vol. 99, pp. 220-
        229, 2015.
        [http://dx.doi.org/10.1016/j.matchar.2014.11.030]

[23]    R. Shashanka, and D. Chaira, "Optimization of milling parameters for the synthesis of nano-structured
        duplex and ferritic stainless steel powders by high energy planetary milling", *Powder Technol.,* vol.
        278, pp. 35-45, 2015.
        [http://dx.doi.org/10.1016/j.powtec.2015.03.007]

[24]    R. Shashanka, D. Chaira, and B.E. Kumara Swamy, "Electrocatalytic Response of Duplex and Yittria
        Dispersed Duplex Stainless Steel Modified Carbon Paste Electrode in Detecting Folic Acid Using
        Cyclic Voltammetry", *Int. J. Electrochem. Sci.,* vol. 10, pp. 5586-5598, 2015.

[25]    S. Gupta, R. Shashanka, and D. Chaira, "Synthesis of nano-structured duplex and ferritic stainless steel
        powders by planetary milling: An experimental and simulation study", *IOP Conf. Series: Materials
        Science and Engineering,* vol. vol. 75, 2015p. 012033
        [http://dx.doi.org/10.1088/1757-899X/75/1/012033]

[26]    R. Shashanka, and D. Chaira, "Effect of sintering temperature and atmosphere on non-lubricated
        sliding wear of nano-yttria dispersed and yttria free duplex and ferritic stainless steel fabricated by
        powder metallurgy", *Tribol. Trans.,* vol. 60, pp. 324-336, 2017.
        [http://dx.doi.org/10.1080/10402004.2016.1168897]

[27]    C. Suryanarayana, "Recent Developments in Mechanical Alloying", *Rev. Adv. Mater. Sci.,* vol. 18, pp.
        203-211, 2008.

[28]    C. Suryanarayana, E. Ivanov, and V.V. Boldyrev, "The science and technology of mechanical
        alloying", *Mater. Sci. Eng. A,* vol. 304–306, pp. 151-153, 2001.
        [http://dx.doi.org/10.1016/S0921-5093(00)01465-9]

[29]    C. Suryanarayana, "Mechanical alloying and milling", *Prog. Mater. Sci.,* vol. 46, pp. 1-184, 2001.
        [http://dx.doi.org/10.1016/S0079-6425(99)00010-9]

[30]    K & Us Equipment, "SPEX 8000D Dual Mixer/Mill Grinder, Pulverizer High-Energy Ball Mill,
        Shaker Mill",  https://www.kandus.com/Configurations/39593SPEX8kD.html

[31]    S. Qiao, "Working Principle Of Lab Planetary Ball Mill", *Changsha Deco Equipment Co..*
        https://www.deco-ballmill.com/info/working-principle-of-lab-planetary-ball-mill-26785144.html

[32]    E. Ajayi, A. Hamweendo, and I. Botef, "Fabrication of Copper/reinforced Carbon Nanotube Bipolar
        Plate using Cold Spray Technique", *International Conference on Research Innovations in Science and
        Engineering (RISE'2016),* 2016 Mauritius

[33]   Mikrons, "How an Attritor Works?", http://www.attritor.in/attritor_working.html

[34]   Turbosquid, https://www.turbosquid.com/3d-models/ 3d-model-lathe-machine-torna/1087299

[35]   MNB Precision Ltd, "CNC Machining Services", https://www.mnbprecision.com/cnc-machinin--services/

[36]   E.P. Maynard, "Mixing & Blending", https://www.powderbulksolids.com/article/mixing-blending-1

[37]   NPTEL, Lecture 10, "Powder Metallurgy", https://nptel.ac.in/courses/Webcourse-contents/II--ROORKEE/MANUFACTURINGPROCESSES/Metal%20Forming%20&%20Powder%20metallurgy /lecture10/lecture10.htm

[38]   M. Russo, "Understanding the Powdered Metal Process: Spotlight on Compacting",

[39]   Inovar Communications Ltd, "Sintering in the Powder Metallurgy Process, Powder metallurgy review", https://www.pm-review.com/introduction-to-powder-metallurgy/sintering-in-the-powder -metallurgy-process/

[40]   V.L. Vilesh, "Sintering - Ceramic fabrication, Science", https://www.slideshare.net/VILESHVL /sintering-57885763

[41]   R.M. German, P. Suri, and S.J. Park, "Review: liquid phase sintering", *J. Mater. Sci.,* vol. 44, pp. 1-39, 2009.
[http://dx.doi.org/10.1007/s10853-008-3008-0]

[42]   GKN Powder Metallurgy, PM expertise drives automotive electrification, https://www.gknpm.com/en/innovation/electrification/

[43]   M. Shafi, *Powder Metallurgy-Present and Future Scope,* 2016.

[44]   H. Wang, X. Li, Z. Wang, and R. Xu, "Influence of Electroplated CBN Wheel Wear on Grinding Surface Morphology of Powder Metallurgy Superalloy FGH96", *Materials (Basel),* vol. 13, no. 4, 2020.E1005
[http://dx.doi.org/10.3390/ma13041005] [PMID: 32102253]

[45]   3C Interglobal - Western Managed Manufacturer - World Class Design and Production, http://www.3cinterglobal.com/Plastics.htm

[46]   Meyersintermetall, Soft magnetic materials, https://www.sintermetall.ch/en/products/soft-magneti--materials#:~:text=Sintered%20soft%20magnetic% 20materials%20are,FeCo

[47]   AMES sintered components manufacturer, https://www.ames-sintering.com/soft-magnetic-parts/

[48]   Edurite, "Advantages, and Disadvantages of Powder Metallurgy", http://www.edurite.com/kbase/advantages-and-disadvantages -of-powder-metallurgy

# CHAPTER 7

# Corrosion

**Abstract:** In this chapter, we have discussed the basic principles of corrosion, factors affecting corrosion, effect of corrosion, types of corrosion, mechanism of corrosion, how to control corrosion, anodic and cathodic coating. This chapter enables the student to learn the fundamentals of corrosion and its effects, disadvantages of corrosion. Corrosion is the process of deterioration of metals as well as alloys in the presence of an electrochemical or chemical environment. It is an irreversible chemical reaction that results in the dissolution of the material. It also deals with the small visual change to complete failure of giant technical systems and causes a big blow to the economy and even hazard to people.

**Keywords:** Anodic and cathodic coatings, Anode, Cathode, Coating, Corrosion, Corrosion cell, Consequences of corrosion, Corrosion inhibitors, Crevice, Differential aeration, Dry and wet corrosion, Economic effects, Electrolyte, Electro plating, Electroless plating, Health effects, Galvanic, Pilling bedworth rule, Pitting corrosion, Waterline.

## 1. INTRODUCTION

Corrosion plays a negative role and affects our society in a significant way. The main victims of corrosion are bridges, electrical towers, automobiles, our daily use metal parts, ships, buildings, appliances, and energy distribution systems, *etc.,* [1]. We have seen a brown color coating on old metal pipes, old metal gates, old ships and on old automobile parts; then what is this brown coating? This is nothing but "Rust" and the phenomenon is known as corrosion. Sometimes, the oxide form of metals is also considered as corrosion. Fig. (**1**) shows the corrosion of metals and alloys [1]. Most of them can readily undergo corrosion and some of them do not. What is the reason for that? Materials like stainless steel, copper do not undergo corrosion easily due to the formation of metal oxide and this metal oxide coating acts as a protective layer and protects from corrosion. The corrosion destroys the metallic surfaces and obtrudes the working of the equipment and results in a massive economic loss to industries. Hence it is very essential to study the reason for corrosion, types and the effect of corrosion [2].

**Shashanka Rajendrachari & Orhan Uzun**

**Fig. (1).** Corrosion of metals and alloys [1].

## 2. CAUSES OF CORROSION

Corrosion generally takes place due to the presence of water, oxygen, excess salts, excess load, bacteria, electric current, *etc*. Below are some of the common reasons for corrosion:

**Water Characteristics:** The physico-chemical properties of water such as basicity, acidity, hardness, presence of anions, the temperature of the water, pH and dissolved salts cause corrosion. The deviation in the above properties can significantly affect the corrosion rate. Excess carbon dioxide and oxygen, low pH and alkalinity, non-carbonate hardness, and more amount of dissolved solids can increase the rate of corrosion [2].

**Bacteria:** Bacteria are another reason for corrosion and it is quite difficult to believe but it is true. The rate of corrosion increases with tremendous speed due to the piling up of bacterial clusters. This enhances the oxygen concentration in the cell and causes pitting and tuberculation. These clusters of bacteria can also produce $CO_2$ and block the deposition of $CaCO_3$ scale [2].

**Other Factors:** The rate of corrosion also affects the characteristics, composition of the metals and alloys. The rate of corrosion depends upon the ground boundary impurities, cyclic loading, and electrical currents.

## 3. WHAT IS CORROSION?

Corrosion can be considered a powerful universal problem, and it is everywhere [3]. Most of the metals are found in nature as ores, especially in their oxide forms. These ores are extracted from the earth's crust to convert into metals by applying some amount of energy. Similarly, the pure metals can be converted into their oxide by releasing the energy and this process is called corrosion [4 - 6]. The entire process is represented in Fig. (2) [6].

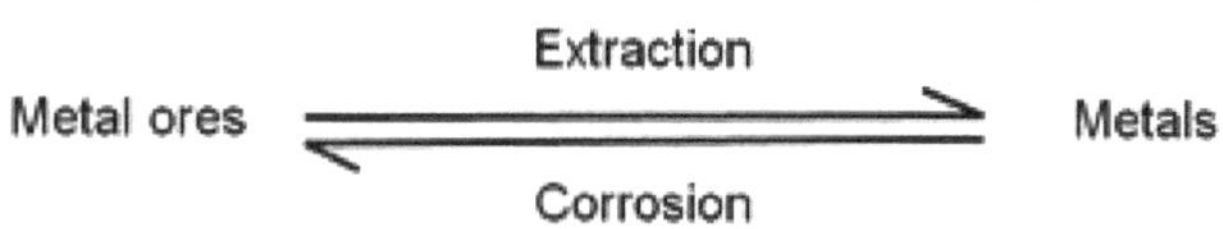


**Fig. (2).** Representation of corrosion [6].

## 4. OUTCOME OF CORROSION

The outcomes of corrosion are many and inevitable and are listed below;

- Economic loss [7]
- Health effects [7]
- Technological effects [7]
- Cultural effects [7]
- Safety effects [7]

## 5. ECONOMIC LOSS

One of the major consequences of corrosion is a major economic loss to industries, and the nation. The economic blow to industries is due to reduced productivity, time and energy loss, *etc.* The major victims of corrosion are pipelines in oil and gas industries, electrical power plants and chemical processing plants, thereby ultimately resulting in shutdown or reduced efficiency [7]. Corrosion also affects our day to day life by decreasing the efficiency of water heaters by forming a scale coating. It also results in changing the taste of water to a bitter taste [7].

## 6. HEALTH EFFECTS

In recent years, technology has brought revolutionary changes in our health systems due to the increasing use of metal prosthetics like pacemakers, pins, plates, hip joints, and other implants. Even though these prosthetics are made up of new types of alloys but cannot control corrosion. They can easily undergo corrosion in the presence of body fluids, oxygen and result in the broken connections in pacemakers, inflammation in tissue around the implants, and fracture of implants [7].

## 7. TECHNOLOGICAL EFFECTS

We have witnessed a lot of super technological inventions but the development of new technology is slightly hindered by corrosion problems because materials are required to withstand higher temperatures, higher pressures and more intense corrosive environments altogether. The technology of drilling for oil in sea and land has also been affected by corrosion, thereby reducing our progress in the energy sector. Therefore, corrosion always prevents the development of technologically workable systems [7].

## 8. CULTURAL EFFECTS

Corrosion deteriorates the precious artifacts, bridges, palaces of our ancestors. The modern-day pollution and corrosion are affecting our valuable cultural heritage [7].

## 9. SAFETY EFFECTS

The biggest safety concerns are corroded structures like bridges, big equipment in industries. Corrosion can occur anytime and anywhere and questions the safety of our modern-day structures like aircraft, bridges, automobiles, pipelines and so on. Even small negligence of corrosion can cause loss of lives, resources, and properties [7].

## 10. EFFECT OF CORROSION ON METALS

Corrosion is a serious problem and needs to be controlled; otherwise, most of the metals will undergo corrosion very easily and lose their engineering properties like strength, hardness, wear resistance, *etc*. Corrosion affects the quality of the materials from economic and environmental points of view; below are some of the effects of corrosion:

1. Loss of metal.
2. Loss of time.

3. Reduction in value of goods.
4. Contamination of fluids due to the chemical reaction.
5. Changes in surface properties.
6. Mechanical damage.

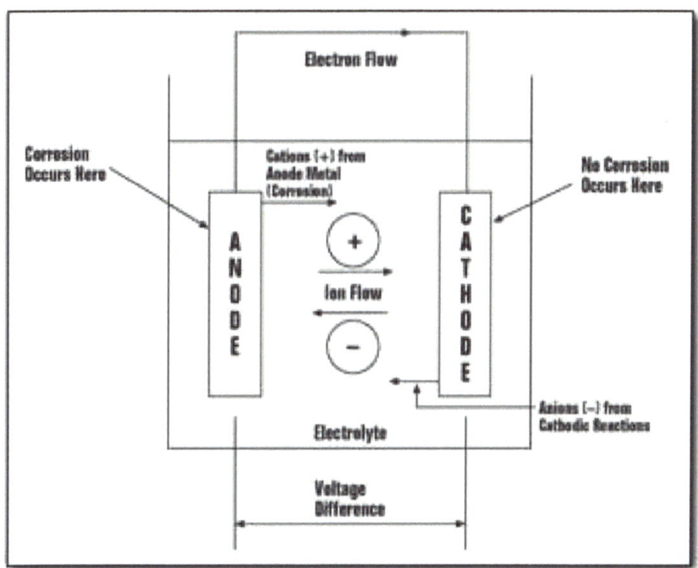


**Fig. (3).** Corrosion cell [8].

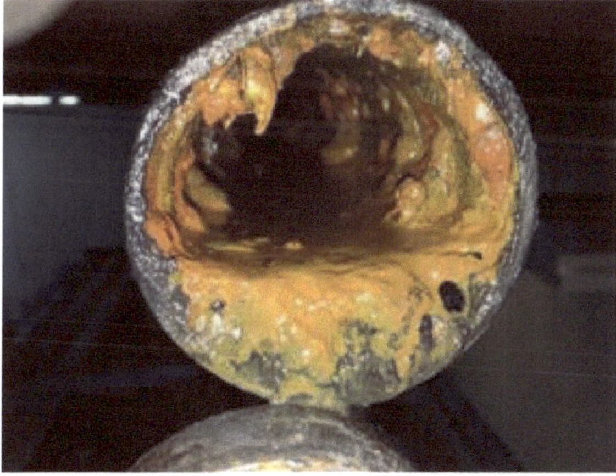

**Fig. (4).** Liquid metal corrosion [10].

**Fig. (5).** Wet or electrochemical corrosion of ship in the ocean [11].

**Fig. (6).** Galvanic corrosion of steel nut and copper bolt [13].

## 11. CONDITIONS NECESSARY FOR CORROSION

### 10.1. Corrosion Cell

As discussed earlier, corrosion is an electrochemical process and it occurs due to the formation of electrochemical cells. Mainly four components are required for the corrosion to take place and they are listed below.

1. **Anode:** The metal electrode where oxidation occurs (loss of electrons).

2. **Cathode:** The metal electrode where reduction occurs (gain of electrons).

3. **Electrolyte:** The electrically conductive medium containing anode and cathode. It should help in the easy movement of ions.

4. **Electrical Connection:** It carries the electrons from the anode to the cathode.

If anyone of the above component is missing from the corrosion cell, then corrosion will not proceed further and it will stop.

In a corrosion cell, the anodic reaction involves the release of cations and electrons. The metal cations and the electrons formed during oxidation (at the anode) begin to migrate from anode to cathode through an electrolyte and electrical connections, respectively, as shown in Fig. (3) [8]. At the cathode, a reduction reaction will take place; therefore, it remains uncorroded. The electrode where oxidation (anode) takes place will always undergo corrosion. On the other hand, the electrode where the reduction (cathode) reaction takes place will not undergo corrosion. Corrosion takes place only when a voltage difference exists between an anode and the cathode. As the potential difference between anode and cathode increases, the rate of corrosion also increases.

## 12. TYPES OF CORROSION

All metallic atoms consist of valence electrons which can be either donated (ionic bond) or shared (covalent bond). As discussed earlier, during corrosion, the metallic materials get ionized (oxidation) and release the electrons. The movement of electrons results in setting up a galvanic or electrochemical cell, and therefore, oxidation or reduction reactions take place.

Corrosion takes place due to many factors and the type of corrosion depends upon the temperature, oxygen, humidity, type of metals used, electrolyte, *etc.* Therefore, corrosion is broadly classified as follows:

1. Dry corrosion or chemical corrosion.
2. Wet corrosion or electrochemical corrosion.

### 12.1. Dry Corrosion

It is a type of corrosion in which the metal is converted into its oxide when atmospheric gases like $O_2$, $SO_2$ and $CO_2$ attack directly on metal in the absence of moisture or liquid phase or conducting liquids. There are mainly three types of dry corrosion and they are discussed below:

### 12.1.1. Oxidation Corrosion

It is a type of dry corrosion and it involves a direct attack of oxygen on a metal surface at room temperature in the absence of moisture.

#### 12.1.1.1. Mechanism

When a metal is exposed to oxygen or air, it gets oxidized and loses its valence electrons; then reduction of oxygen takes place and metal starts corroding.

Anodic reaction: $2M \rightarrow 2M^{n+} + 2n^{e-}$

Cathodic reaction: $\frac{n}{2}O_2 + 2n^{e-} \rightarrow nO^{2-}$

Overall reaction: $2M + \frac{n}{2}O_2 \rightarrow 2M^{n+} + nO^{2-}$

Where, $2M^{n+}$ is a metal ion and $nO^{2-}$ is oxide ion, respectively. These formed ions are responsible for the formation of a protective layer on the top of the metals and restrict further corrosion of metals. The size of the cation ($M^{n+}$) is smaller than anion ($O^{2-}$), and therefore, cation will diffuse much effectively than anion through the protective layer for further continuation of oxidation. The diffusion of cation takes place only when the protective oxide layer is sufficiently porous. Hence oxidation corrosion mainly depends upon the characteristics of the oxide layer or barrier film.

- The stronger the oxide layer, the lesser is the corrosion. It acts as a protective layer.

- If the oxide layer is unstable, it decomposes to give back metal and oxygen and does not undergo oxidation corrosion.

- The more the volatile nature of the oxide layer, the higher the rate of oxidation corrosion.

- The greater the porosity in the oxide layer, the faster is the oxidation corrosion.

#### 12.1.1.2. Pilling Bedworth Rule

As the specific volume ratio increases, the rate of corrosion decreases. If the volume of the metal oxide layer is more than the volume of metal from which it is formed, then the non-porous layer strongly adheres to the base metal and acts as a protective layer [9].

$$Specific\ volume\ ratio = \frac{Volume\ of\ metal\ oxide\ formed}{Volume\ of\ the\ metal}$$

As the volume of metal exceeds the volume of the metal oxide, then the corrosion takes place very rapidly. This results in the formation of a non-continuous, non-protective and more porous oxide layer followed by cracks and pores. This results in easy access to atmospheric oxygen to diffuse into the underlying metal, causing corrosion.

Examples: Lithium, sodium and potassium.

### 12.1.2. Liquid Metal Corrosion

When a molten metal is passed through solid metal at high temperature, then part of the solid metal undergoes dissolution or internal penetration and this causes corrosion called liquid metal corrosion. Fig. (**4**) depicts the liquid metal corrosion inside a pipe [10].

Examples: Sodium metal used in a nuclear reactor as coolant can lead to the corrosion of cadmium. Liquid metal mercury dissolves most of the metals by forming their amalgams, thereby corroding them.

### 12.1.3. Corrosion by other Gases

Some gases like $Cl_2$, $SO_2$, $H_2S$, and NOx react with certain metals and form corrosion products, which may be protective or non-protective.

Dry chlorine ($Cl_2$) reacts with silver (Ag) and forms silver chloride (AgCl), which acts as a protective layer. On the other hand, the $SnCl_4$ layer is volatile and it is non-protective [9]. In petroleum industries, $H_2S$ attacks steel at high temperatures and forms the FeS layer, which is porous and causes corrosion.

The rate of corrosion mainly depends upon the nature of the environment and the characteristic of the oxide film formed.

## 12.2. Wet Corrosion

Wet corrosion is a common type of corrosion that usually takes place in the presence of moisture by forming the electrochemical (galvanic) cells. Wet corrosion occurs

1. When a metal is in contact with conducting liquid.

2. When two dissimilar metals are partially immersed in a solution.

As discussed earlier, at anode, oxidation reaction (loss of electrons) takes place and at cathode, reduction reaction (gain of electrons) takes place. Therefore, anodic metal undergoes deterioration (corrosion) by either dissolving or assuming a combined state. Only anodic materials undergo corrosion but cathodic materials remain uncorroded [11].

Anodic reaction: $M \rightarrow M^{n+} + n^{e-}$

During the anodic reaction, the metal undergoes oxidation to form metal ions ($M^{n+}$) and electrons. Metal ions dissolve in solution or form compounds such as oxide.

A cathodic reaction does not affect the cathode; therefore most of the metals cannot be reduced further. At cathode, dissolved constituents in the conducting medium accept the electrons from the anode to form some ions like $OH^-$, $O^{2-}$. Fig. (**5**) shows the impact of wet or electrochemical corrosion [11].

## 13. DIFFERENCES BETWEEN DRY AND WET CORROSION

**Table 1. Comparision of dry and wet corrosion [6].**

| Dry Corrosion | Wet Corrosion |
|---|---|
| Involves direct attack of atmospheric gases on metal in the absence of moisture liquid phase | Involves an electrochemical attack on metals in aqueous environments |
| Less prevalent | More prevalent |
| The corrosive media include vapors, gases, *etc.* | The corrosive media occur at liquid (aqueous) phase |
| It is of various types: Oxidation corrosion, corrosion by other gases such as $Cl_2$, $SO_2$, $H_2S$, $NO_x$ and liquid metal corrosion | It is of mainly four types: differential metal corrosion, differential aeration corrosion, waterline and pitting corrosion |

## 14. TYPES OF WET OR ELECTROCHEMICAL CORROSION

Wet corrosion is broadly classified into:

1. Differential metal corrosion (Galvanic corrosion)
2. Differential aeration corrosion (Concentration cell corrosion)
3. Waterline corrosion
4. Crevice corrosion
5. Pitting corrosion

## 14.1. Differential Metal Corrosion (Galvanic Corrosion)

Differential metal corrosion is one of the more prevalent wet corrosions. It occurs when two dissimilar or different metals are in contact with each other through an electrolyte; then, the metal with high electrode potential acts as a cathode and the metal with lower electrode potential acts as an anode. As discussed earlier, only anodic materials undergo corrosion and the cathode remains uncorroded. This potential difference between 2 dissimilar metals produces an electron flow between them and more active metal (low electrode potential) undergoes corrosion. This form of corrosion is called galvanic or differential metal corrosion. The rate of galvanic corrosion mainly depends upon the types of metals, potential difference, the relative size of anode, pH, temperature, humidity, *etc.*

Some of the examples of differential metal corrosion (galvanic corrosion) are:

- Buried iron pipeline connected to the zinc bar
- Steel pipe connected to copper plumbing
- Corrosion between the outer copper skin and the wrought iron support structure of Statue of Liberty [12]
- Steel propeller shaft in bronze bearing
- Zinc coating on mild steel
- Lead–in solder around copper wires
- Steel nut with copper bolt (see Fig. **6**) [13]

Preventive Measures:

- Keeping a thin layer of an insulator between two metals or materials
- Selecting materials having a very less potential difference
- Avoiding the contact of metals with conducting medium
- Coating

## 14.2. Differential Aeration Corrosion (Concentration Cell Corrosion)

It occurs when the metal surface is exposed to the different air or oxygen concentrations; the metal part exposed to a low concentration of oxygen acts as an anode and undergoes corrosion. Whereas, the other metal part exposed to higher oxygen concentration acts as a cathode and does not undergo corrosion. Due to the differences in the oxygen level in the metal, aeration corrosion is the result.

Fig. (7) shows a metal strip of iron partially immersed in an aerated solution of NaCl [15]. The part of the metal strip outside the solution of NaCl acts as a cathode due to the availability of a high concentration of oxygen and does not undergo corrosion [14, 15]. Generally, the cathodic reaction takes place in the presence of oxygen only. But, the metal part inside the NaCl solution acts as an anode due to the less availability of oxygen and undergoes aeration corrosion.

At anode: Fe Fe$^{2+}$ + 2e$^-$

At cathode (near water line): O$_2$ + 2H$_2$O + 4e$^-$ 4OH$^-$

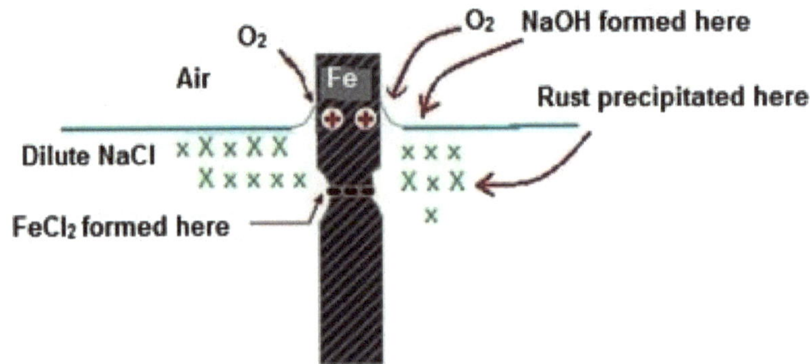


**Fig. (7).**  Differential aeration corrosion [15].

The examples of differential aeration corrosion are

- Half-immersed iron plate in aqueous solution
- Steel pipe carrying any liquid exposed to the atmosphere
- Ocean-going ships
- Metal storage tanks
- Steel pillar of the bridge immersed in the water

Preventive Measures:

- Using metallic coating, electroless plating or chemical conversion methods
- Maintaining the material's protective film
- Using inhibitors
- Trying to maintain the same amount of oxygen

## 14.3. Waterline Corrosion

This type of corrosion is almost the same as that of differential aeration corrosion. Waterline corrosion is commonly observed in steel water tanks, ocean-going ships, *etc.* in which portion of the metal is always underwater. The part of the metal below the water is exposed only to dissolved oxygen (acts as the anode), while the part above the water is exposed to a higher concentration of oxygen (acts as the cathode). Thus the metal part below the water line undergoes corrosion; whereas, the metal part above the waterline remains uncorroded. Overall, the metal less exposed to oxygen acts as an anode and corrodes. Whereas, the metal exposed to a high concentration of oxygen acts as an anode and remains uncorroded, as shown in Fig. (8) [6]. A high concentration of oxygen is required to undergo a cathodic reaction. This is called water line corrosion.

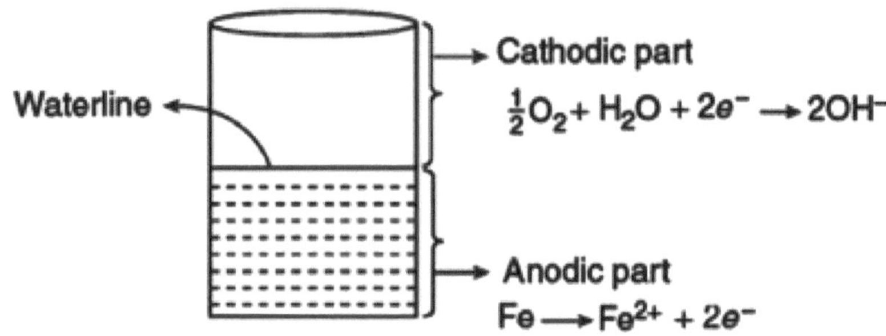


**Fig. (8).** Waterline corrosion [6].

## 14.4. Crevice Corrosion

Crevice corrosion is a localized form of corrosion and usually takes place due to the stagnant microenvironment in which there is a difference in the concentration of ions between two areas of a metal [16]. Crevice corrosion occurs in shielded areas such as those under washers, bolt heads, gaskets, *etc.*, where oxygen is restricted. Generally, crevice corrosion takes place due to design failure.

Preventive Measures:

- Using anti-corrosive metals like chromium, molybdenum, *etc.* in high concentration while preparing alloys
- Careful design and fabrication to avoid crevices
- By periodic inspection and closing the crevices by continuous welding, caulking or soldering
- Proper use of sealants and protective coatings

## 14.5. Pitting Corrosion

Pitting is one of the most common and destructive types of corrosion. It is hard to predict, detect and to characterize; therefore, most of the corrosion researchers or scientists are performing more and more research on pitting corrosion. Pitting corrosion is a localized form of corrosion, in which either a local anodic point or, more commonly, a cathodic point forms a small corrosion cell with the surrounding normal surface [16]. In other words, pitting corrosion is also defined as a local break in the passive layer of the stainless steel provoked by an electrolyte rich in chloride and or by sulfides. The mechanism of pitting corrosion is depicted in Fig. (9) [17].

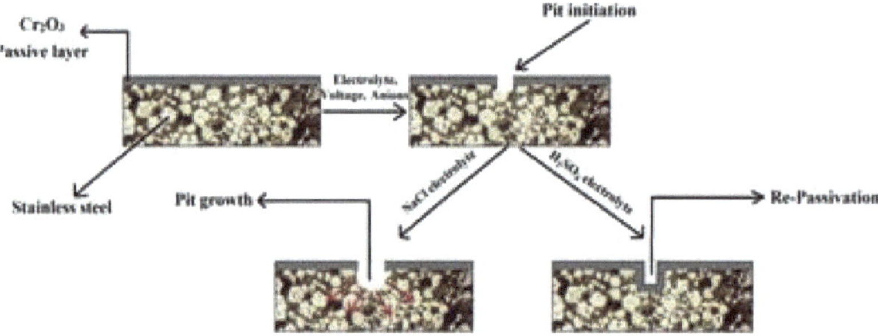


**Fig. (9).** Mechanism of pitting corrosion [17].

The important steps involved in the pitting process are the breakdown of the passive film, metastable pitting and pit growth. The mechanism of pitting corrosion involves the dissolution of the protective passive film and gradual acidification of the electrolyte caused by its insufficient aeration [17]. This increases the pH of the pits by increasing the anion concentration. This type of localized pitting corrosion results in an accelerated failure of structural components by pitting or by acting as an initiation site for cracking. It is also caused by a non-uniform surface in the metal structure itself. Pitting is dangerous because it can lead to failure of the structure with a loss of metal.

## 15. OTHER TYPES OF CORROSION

### 15.1. Stress Corrosion Cracking

Stress corrosion cracking is a combination of both mechanical and electrochemical corrosion processes that results in cracking of certain materials. The impact of this corrosion is very high and results in an unexpected sudden brittle failure. Generally, stress corrosion occurs in ductile metals subjected to stress levels well below their yield strength. Internal stresses in a material can be sufficient to initiate an attack of stress corrosion cracking [18]. In the beginning, a small pit is formed and develops into a crack due to applied or residual stress in the material and this crack formation results in a new active metal surface. The new metal surface again undergoes corrosion due to the cyclic loading. This results in further crack propagation and leads to the exposure of new highly active metal surfaces in the crack.

## 15.2. Intergranular Corrosion

However, when the materials are cooled slowly from high temperatures or even heated between 425 and 815°C, chromium carbides precipitate at the grain boundaries. These carbides have higher chromium content in comparison to the matrix. Inter-granular corrosion takes place rapidly in the de-chromed sites. Fig. (**10**) depicts the Intergranular corrosion [19].

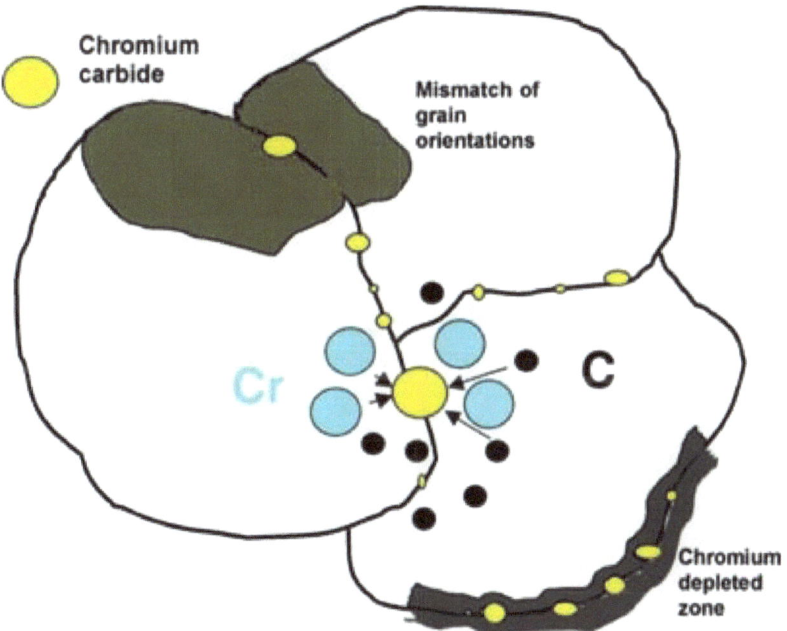


**Fig. (10).** Intergranular corrosion [19].

Intergranular corrosion can be prevented by quenching (rapid cooling) the materials at 1050/1100°C.

## 15.3. Microbiological Corrosion

Microbiological corrosion involves the degradation of materials by bacteria, molds, and fungi or their byproducts. These microbes can be aerobic or anaerobic. Aerobic bacteria decrease the concentration of oxygen in the medium and cause corrosion. Bacterial corrosion may appear in the form of pitting corrosion at pipelines of the oil and gas industry. Anaerobic bacteria like Desulfovibrio and Desulfotomaculum are responsible for the layers of metal sulfides and hydrogen sulfide to smell [20].

## 15.4. Soil Corrosion

Buried pipes, cables, tank bottoms, *etc.*, undergo corrosion due to moisture, pH of the soil, ionic species like chlorides and micro-organisms like bacteria. It is further enhanced by the differential aeration of various parts of the soil. Fig. (**11**) depicts soil corrosion of pipes buried inside the earth [21].

**Fig. (11).** Soil corrosion of pipes buried inside the earth [21].

## 15.5. Erosion Corrosion

Due to mechanical wear and tear, there is a relative motion of the corrosive liquid over the metal surface. This motion increases the rate of corrosion on the surface of a metal and is called erosion-corrosion. It is most common in copper tubes.

Preventive measures [22]:

- streamline the piping to reduce turbulence
- control fluid velocity
- using corrosion inhibitors or cathodic protection to minimize erosion-corrosion

## 16. FACTORS INFLUENCING THE RATE OF CORROSION

The rate of corrosion mainly depends upon following primary and secondary factors mentioned below:

### 16.1. Primary Factors

1. Nature of the metal
2. Electrode potential difference
3. The surface state of the metal
4. Hydrogen overvoltage
5. Formation of protective films by metals

### 16.2. Secondary Factors

1. pH
2. Temperature
3. Anodic and cathodic area effect
4. Polarization of electrodes
5. Conductance of the medium
6. Humidity

## 17. METHODS TO CONTROL CORROSION

### 17.1. Application of Protective Coatings

Metallic structures can be protected from corrosion in many ways, but a common method is the application of protective coatings. These protective coatings are generally made from plastics, polymers, paints and coating the layer of noble metals on the metallic structures. These coatings act as an impermeable barrier between the metal and the oxidant, thereby protecting the metallic structures from further corrosion. The protective coating should be uniform throughout the surface, completely covering the entire metallic structures and should be flawless, or else it will initiate corrosion and the rate of corrosion will increase as the time will pass, ultimately damaging the metallic structures.

**Metal Coating:** Deposition of a protective metal over the surface of a base metal.

**Anodic Coating:** It is produced by coating the base metal with more active metals (Zn, Mg, Al) which are anodic to the base metal. Even if the anodic coating is ruptured, the base metal is protected. The exposed surface of the metal is cathodic to the coating metal and the coating metal undergoes corrosion. Here, coated metal sacrifices itself by protecting the base metal; hence it is also called sacrificial coating.

Example: Galvanization of iron and steel objects

**Cathodic Coating:** It involves coating a base metal with more noble metals that are cathodic to the base metal (Cu, Ni, Ag). Coating metals are less reactive than the base metal and are less susceptible to corrosion. It is protective only when the coating is undamaged, non-porous and continuous. Intense localized corrosion occurs when it is discontinuous (setting up of galvanic cell).

Example: Tinning

**Inorganic Coating:** Inorganic coatings are generally chemical conversion coatings. A surface layer of the metal is converted into a compound by chemical or electrochemical reactions, which form a barrier between the underlying metal surface and the corrosion environment.

Example: phosphate coating

**Organic Coating:** In this type, paints and lacquers are the most widely used anti-corrosion coatings.


## 18. POLARIZE OR SHIFT THE POTENTIAL OF THE METAL

This is another method of preventing the corrosion and this method involves polarizing or shifting the potential of the metal and forcing an anode to act as a cathode in an electrochemical cell. By changing the potential, one can reform an anode to act as a cathode, thereby protecting the anodic materials. The galvanization of steel with a coating of zinc is an example of this method of corrosion control.


## 19. MATERIALS SELECTION

Material selection plays an important role in eliminating corrosion. It is always recommended to use corrosion-resistant metals like chromium, molybdenum, *etc.* in alloys. Below are some of the basic recommendations of materials selection to eliminate corrosion:

a. It is recommended to use corrosion-resistant materials like plastics, polymers, and non-metallic materials in corrosion friendly environments.
b. Avoid using dissimilar metals having a large potential difference.
c. Avoid the use of alloys susceptible to stress corrosion.
d. It is recommended to use defect-free (good designed) structures to avoid crevice corrosion.
e. Use alloys which are stable even at higher temperatures.
f. Use composite materials, especially ceramic and polymer matrixes.

## 20. CORROSION INHIBITORS

Corrosion inhibitors can be added in small amounts to solutions in contact with metals. Inhibitors can prevent either the anodic or the cathodic reaction of corrosion cells by forming insoluble films over the anode or cathode sites of the cell. Hydrogen liberation can be retarded or prevented by diffusion of hydrogen ions to the cathode or by increasing the hydrogen overvoltage on the metal surface. Oxygen absorption can be done by adding scavengers like hydrazine and sodium sulfite. These are substances added in small concentrations to a corrosive environment to decrease the corrosion rate. There are mainly two types of corrosion inhibitors:

**Anodic Inhibitors:** Examples of anodic inhibitors include chromate, tungstate, molybdate, *etc.*

**Cathodic Inhibitors:** Generally, two types of cathodic reactions take place; liberation of hydrogen in acidic solutions and -OH ions in alkaline solution or absorption of oxygen. These can be of two types.

Organic cathodic inhibitors such as amines, mercaptans, thioureas, sulphoxides form a protective layer on cathodic regions, prevent the evolution of hydrogen, thus preventing the corrosion.

Inorganic cathodic inhibitors such as sulfates of Mg, Mn, Ni, and Zn are used in a neutral or alkaline medium. These inhibitors react with OH ions liberated at the cathode and form insoluble hydroxides, which form protective film over cathode areas and prevent corrosion.

## 21. ELECTROPLATING

Electroplating is a process of depositing a metal on the substrate by an electrochemical process, and it involves passing an electric current through an electrolyte. This can be achieved by dipping two electrodes (anode and cathode) into the electrolyte and connecting them into a circuit with a battery or other power supply [23]. The electrolyte should be chosen in such a way that it must contain the electrode's metal ion.

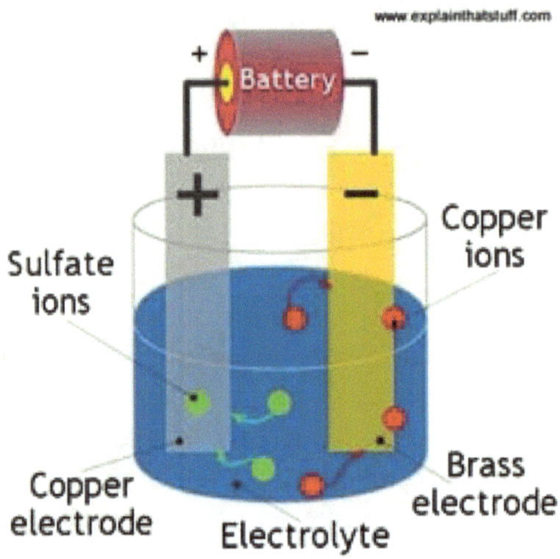


**Fig. (12).** Electroplating instrumental setup [23].

Once we switch on the power supply, the electricity begins to flow and it dissociates the electrolyte into metal ions and these metal ions are deposited in the form of thin layer on the cathode or the material to be electroplated. Metals like gold, silver, zinc, copper, tin, cadmium, chromium, nickel, platinum and lead can be electroplated. Fig. (**12**) shows the representation of the electroplating instrumental setup.

### 21.1. Plating Bath

An electrolyte used in electroplating should contain some of the below characteristics:

1. It should provide a source of the metal or metals being deposited.
2. It must form complexes with ions of the deposition metal.
3. It should provide conductivity and stabilize the solution against hydrolysis and other side reactions.

4. It should act as a buffer to stabilize the sudden variation in pH.
5. Modify other properties either of the solution or of the deposit for specific cases.
6. It should help in the dissolution of the anode.

The purposes of electroplating an article:

- To improve the appearance.
- To increase the protection (long-lasting).
- To impart special surface properties.
- To enhance the mechanical properties.

The applications of electroplating are as follows:

i. Steel parts in automobiles are plated with chromium to enhance the corrosion resistance properties.
ii. Gold plated jewelry.
iii. Steel and aluminum components are coated with nickel or chromium or brass to improve their appearance.
iv. Coating silver on copper or brass materials to enhance the superconductivity.
v. Electroplating can also be used to reduce friction, abrasion, and protection against radiation.

## 21.2. Electroless Plating

Electroless plating is also known as autocatalytic plating and it is a type of plating that deposits certain metals or materials without using an external power source. It is based on the catalytic reduction of metal ions on the surface of the substrate that is being plated. For example, depositing nickel on the metal or material parts in an aqueous solution by creating a catalytic reduction of nickel ions to plate the part without any electrical energy. Metal can be deposited even along edges, inside holes and over irregularly shaped objects.

### *21.2.1. Applications of Electroless Plating*

To beautify the product by imparting more corrosion resistance properties, electroless plating methods are often a better choice due to very hard and less porous deposition. Following are some of the applications of electroless plating in various industries [24],

1. In food industries, the molds and food processing machine parts are often plated with electroless plating.

2. The valves, balls, and plugs to barrels, pipe fittings are electroless plated in oil and gas industries.
3. Automotive industries use electroless deposited car parts like shock absorbers, cylinders, brake pistons, and gears.
4. Aerospace industries use electroless plated valves, pistons, pumps, rocket components.
5. In chemical industries, the pumps, mixing blades, heat exchangers, filter units are often electroless plated.
6. Molds, dyes, machine parts, spinnerets, extruders in the plastics and textile industries are made up of electroless plating.

## CONCLUSION

The deterioration of a material through a chemical or electrochemical reaction with its surrounding environment is called corrosion. It is a natural phenomenon and has a very dangerous impact on the economy and health of the society. Therefore, we need to control the corrosion of metals and alloys. This chapter describes the basic concepts of corrosion like types, effects, and control of corrosion. Corrosion scientists all over the world are trying to reduce the rate of corrosion by using different types of anti-corrosion metals and alloys.

## QUESTIONS

1) What is corrosion?

2) What are the causes of corrosion?

3) Explain in detail the consequences of corrosion.

4) What are the effects of corrosion on metals?

5) What is a corrosion cell? Mention the four important components of the corrosion cell.

6) Explain the importance of components of the corrosion cell in detail.

7) What is dry or chemical corrosion? Mention its different types.

8) Explain the mechanism of oxidative corrosion.

9) What is Pilling Bedworth rule? and why is it necessary?

10) What is wet or electrochemical corrosion? Mention its different types.

11) Explain galvanic or wet corrosion in detail.

12) Differentiate wet and dry corrosion.

13) Explain differential metal corrosion and differential aeration corrosion with examples and its preventive measurements.

14) What is crevice corrosion? How can it be controlled?

15) What is pitting corrosion? Explain its mechanism.

16) Write a note about stress corrosion, intergranular corrosion, microbiological corrosion, soil, and erosion-corrosion.

17) Name some of the factors influencing the rate of corrosion.

18) Explain the methods used to prevent corrosion in detail.

19) What is the anodic and cathodic coating?

20) What are corrosion inhibitors? Explain their different types.

21) What is electroplating? Explain.

22) Mention some of the characteristic properties of an electrolyte used in electroplating.

23) Why electroplating is required?

24) What is electroless plating? Mention their applications.

25) Differentiate electroplating and electroless plating.

## REFERENCES


[1]　　K. Nanan, "The 8 Most Common Forms of Metal Corrosion", *Corrosionpedia,* 2018.

[2]　　TutorVista, "Effects of Corrosion", https://chemistry.tutorvista.com/physical-chemistry/effects-of-corrosion.html

[3]　　K.A. Natarajan, Lecture 1, "Corrosion: Introduction – Definitions and Types, Advances in Corrosion Engineering", *NPTEL Web Course.*

[4]　　M.P. Schultz, "OCE-4518 Protection of Marine Materials Class Notes", *Florida Institute of Technology,* 1997.

[5]　　G.W. Swain, "OCE-4518 Protection of Marine Materials Class Notes", *Florida Institute of Technology,* 1996.

[6]　　Wiley Editorial, Engineering Chemistry, 2nd ed.

[7]　　SCRIBD, "consequences of corrosion", https://www.scribd.com/document/366478737/Consequences-of-Corrosion

[8]　　G.A. Jacobson, "Corrosion - A natural but controllable process", *NACE INTERNATIONAL.*

[9]　M.G.V. Satyanarayana, *Unit - 3: "Corrosion Science.* Engineering Chemistry, 2017.

[10]　Corrosion assessment, ECS Corrosion, https://www.ecscorrosion.com/service/corrosion-assessment/

[11]　Wikimedia, "Wreck of Cabo de Santa Maria",

[12]　Wikipedia, "Galvanic corrosion", https://en.wikipedia.org/wiki/Galvanic_corrosion

[13]　K. Chadchan, "Differential metal corrosion (Galvanic Corrosion)", *chembldeacet,* 2018.

[14]　"Nikhi, Unit 2- Class 3-", *Corrosion,* 2010. [Types of Corrosion].

[15]　I.Z. Awan, and A.Q. Khan, "Corrosion–Occurrence & Prevention", *J. Chem. Soc. Pak.,* vol. 40, pp. 602-655, 2018.

[16]　Gibson Stainless & Specialty Inc, "Corrosion types and prevention",

[17]　R. Shashanka, *Investigation of Electrochemical Pitting Corrosion by Linear Sweep Voltammetry: A Fast and Robust Approach.* Voltammetry, 2018.

[18]　Hilti Corporation, *Corrosion Handbook* Addison-Wesley publishing: Boston, 2015, pp. 1-47.

[19]　The Stainless steel information center, CORROSION, "Intergranular Corrosion", http://www.ssina. com/corrosion/igc.html

[20]　Wikipedia, "Microbial corrosion", https://en.wikipedia.org/wiki/Microbial_corrosion#:~:text =Anaerobic%20corrosion%20is%20evident%20as,low%20mechani cal%20strength%20in%20place

[21]　Sterling Analytical, "Soil Analysis for Corrosion Potential", https://www.sterlinganalytical. com/services/soil-testing/soil-analysis-for-corrosion-potential/

[22]　WebCorr, "Corrosion consulting services, Different Types of Corrosion, Erosion Corrosion", https://www.corrosionclinic.com/types_of_corrosion/Erosion%20Corrosion.htm

[23]　Chris Woodford, Electroplating, https://www.explainthatstuff.com/electroplating.html

[24]　SPC, "Surface Treatment Experts, Electroless plating", https://www.sharrettsplating.com/plating-methods/electroless-plating

CHAPTER 8

# Composites

**Abstract:** In this chapter, we have discussed the basics of composite materials, types, properties, and applications. Students will learn the fundamentals of composite materials and their importance in many engineering fields. Composite materials can be defined as "the combination of a hard and a soft material" or "these are the materials composed of different parts with specific ratios". This chapter comprises of the advantages and disadvantages of composites and also focuses on different fabrication methods of preparing composites. Fundamentals of nanocomposites and their recent improvements are also added in this chapter.

**Keywords:** Ceramic matrix, Composites, Fabrication, Fibers, Fiber-reinforced plastics, Fillers, Laminates, Metal matrix, Nanocomposites, Polymer matrix, Reinforcements, Resin.

## 1. INTRODUCTION

In recent years, composite materials are becoming more popular due to the possibility of fabricating high-tech materials for modern applications. Composites (see Fig. **1**) are revolutionary materials that have been used in various engineering fields for more than 60 years [1, 2]. The important advantages of composite materials over other bulk materials are their high strength to low weight ratios, their stiffness combined with low density [3]. The reinforcing phases are always harder, stronger than the matrix constituent, and impart greater strength and stiffness to the matrix. Composite is a future technology, and therefore, more and more research work is going on all over the world to improve the microstructure, mechanical properties, electrical properties, corrosion resistance properties, and surface properties.

One of the earliest man-made composite materials are bricks made up of straw and mud for constructing buildings during the Egyptian era, as drawn in their tomb paintings [1]. Concrete is another oldest and one of the highly used composite materials in the world.

**Shashanka Rajendrachari & Orhan Uzun**

Nowadays, nanomaterials are impregnated into fibers, and resins are used as a new class of composites. The mechanical, optical, thermal, electrical, biological, chemical, and physical properties of the composites are further improved by reinforcing carbon nanotubes, fullerenes, graphenes, nano-metals, nano-alloys, *etc*.

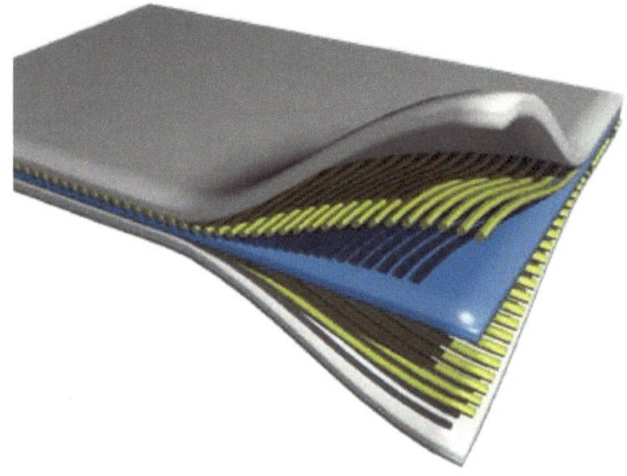

**Fig. (1).** Composite materials [1].

## 2. CLASSIFICATION OF COMPOSITES

Composites are broadly classified into 2 types based on the matrix used and reinforcement form.

Based on matrix constituent:

1. Organic matrix composites (OMCs)
2. Metal matrix composites (MMCs)
3. Ceramic matrix composites (CMCs)

Based on reinforcement form:

1. Fiber-reinforced composites
2. Laminar composites
3. Particulate composites

### 2.1. Based on Matrix Constituent

In these types of composites, the matrix helps in holding the reinforcements intact and distributes the applied stresses to reinforcement materials. Potential matrix materials are those solids that accommodate stress to incorporate other constituents providing stronger bonds for the reinforcing phase [4]. The inorganic matrix materials like polymers, metals, and ceramics are frequently employed to design the composites with great precision. These materials remain elastic under tensile or compressive force before it breaks. Based on matrix constituent, composites can be classified into organic or polymer matrix composites, metal matrix composites, and ceramic matrix composites.

### 2.1.1. Organic Matrix or Polymer Matrix Composites

Polymers are those types of materials that can be prepared very easily and exhibit lightweight, corrosion resistance, high strength to low weight ratios, and desirable mechanical properties. As we know, polymers are mainly of 2 types; thermosets and thermoplastics. Thermosetting plastics are flexible and demonstrate a strong 3D molecular structure that undergoes decomposition after curing instead. By changing the basic composition of the polymers, one can alter the conditions suitable for curing and determine its other characteristics. Therefore, polymers are one of the best and commonly used matrix bases used in the fiber-reinforced plastics, chopped fiber composites form. On the other hand, thermoplastics show 1D or 2D molecular structure and they tend to show an exaggerated melting point at elevated temperatures. Therefore, they can be used as high-temperature resins for aeronautical applications [4]. Thermoplastics, such as polyethylene, polystyrene, polyamides (nylons), polypropylene, *etc.*, can be used as a matrix material in composites. Similarly, thermosetting polymers like epoxy resins, phenol-formaldehyde (Bakelite), urea-formaldehyde, polyester, *etc.*, can be used as matrix materials.

#### 2.1.1.1. Advantages of Polymer Matrix Composites

Due to their lightweight, they are used in aircraft hulls, military and heavy goods vehicles, mass transit systems in trains, subways, and buses. The high chemical corrosion resistance, scratch-resistance, and resilience to seawater made them suitable composites to be used in boats or other marine craft applications.

#### 2.1.1.2. Disadvantages of Polymer Matrix Composites

Generally, most of the polymer matrix composites decompose above 300°C, therefore their service temperatures are limited to below 300°C. Advanced composite structures in military aircraft applications are very expensive and this is a major blow for widespread use in commercial applications. Low production rates are not suitable for high-volume productions.

### 2.1.2. Metal Matrix Composites

In recent years, metal matrix composites are gaining more interest among researchers due to their high strength, high-temperature stability, fracture toughness, and stiffness compared to the organic matrix. Metals can withstand high temperatures very well, even in a corrosive environment than polymer composites. Numerous metals and alloys can be used as matrices, but ultimately the overall strength of the composite materials mainly depends upon the type of reinforcement materials used. Light metals and lightweight alloys are more suitable to be used as matrix materials to fabricate composites that exhibit above mentioned properties. Light metals like titanium, aluminum, and magnesium can be used as a matrix material for aircraft applications. High modulus reinforcements are one of the important requisites of metallic matrix materials to exhibit high strength [4]. The various physical and mechanical properties of the metal matrix composite at different temperatures determine the safe working temperature of composites. Ultimately, the choice of reinforcements plays an important role in deciding the overall properties of metal matrix composites.

### 2.1.3. Ceramic Matrix Composites

Ceramics can be described as crystalline or semi-crystalline inorganic solids, which demonstrate a very strong ionic bonding and a few of them exhibit covalent bonding. Ceramic materials show high melting points, excellent corrosion resistance, high-temperature stability, and high compressive strength. Due to the above properties, they are mainly used for high-temperature operations above 1500°C temperatures. High modulus of elasticity and low tensile strain of ceramic materials made it difficult to add reinforcements to obtain strength improvement. The use of reinforcement with a high modulus of elasticity and also pre-stressing of the fiber in the ceramic matrix may take care of the above problem to some extent. Ceramic materials like alumina and zirconia can be used as matrix materials in fabricating ceramic matrix composites.

## 2.2. Based on Reinforcement Form

The main function of reinforcements is to impart strength and stiffness by acting as an obstacle for the load. Reinforcements can be oriented in all directions to provide better mechanical properties in the direction of the loads imparted on the material. Reinforcements or fibers can be both natural and artificial depending upon the requirement and the extent of stiffness and strength. One of the natural reinforcements is cellulose in wood. But many of the reinforcements, such as glass fiber, carbon fiber, nylons, metals, metal oxides, and ceramics, are man-made. Reinforcements play a crucial role in defining the strength and stiffness of the composites. Based on reinforcement, composites are classified into laminar, particulate, and fiber-reinforced composites.

## 2.2.1. Laminar Composites

These composites are sheets of continuous fiber composites laminated in such a way that each layer has the fibers oriented in a given direction. They include plywood, in which wood and fiber reinforcements are aligned in preferred directions. Layers are alternatively stacked together in such a way that the orientation of the reinforcements varies with each successive layer. This type of layering brings the grain of each layer at right angles to its neighboring layer [5]. The use of fabric material, such as cotton, paper, or glass fiber, dispersed in suitable plastic is some of the examples of these composites. Fig. (2) depicts the laminar composites [5].

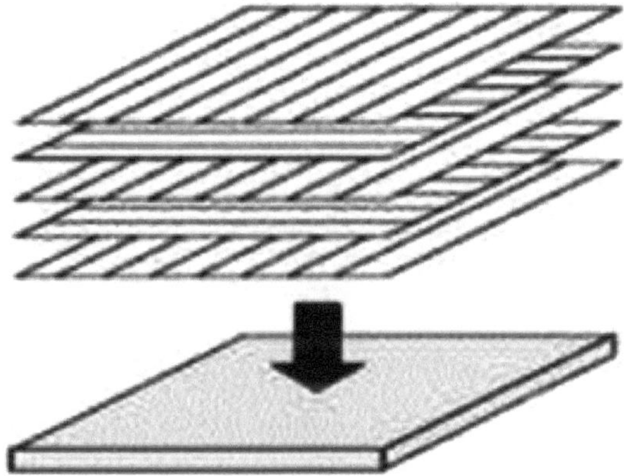

**Fig. (2).** Laminar composites [5].

## 2.2.2. Particulate Composites

One of the best examples of a particulate composite is concrete containing crushed stones or gravel as reinforcements and cement as a matrix. The gravels impart strength while the cement acts as glue and holds the structure together with a strong bond. Particulate composites are prepared from injection molding techniques with various dimensions like very small particles (< 0.25 microns), chopped fibers, platelets, hollow spheres, or new materials such as buckyballs or carbon nano-tubes [6]. Fig. (3) depicts particulate composites [7].

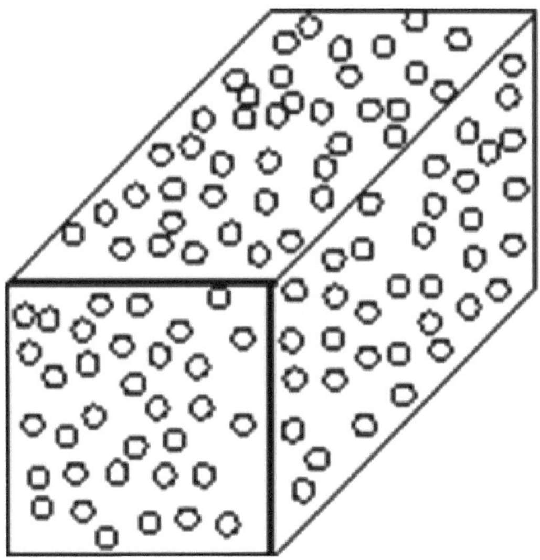

**Fig. (3).** Particulate composites [7].

### *2.2.3. Fiber-reinforced Composites*

These are important and very popular types of composite materials in which fibers act as reinforcement and polymers act as a matrix. These composites show excellent strength and improved mechanical properties due to the reinforcement of strong, stiff, and brittle fibers into a soft, ductile matrix. When an external force is applied, then the matrix material will carry and distribute that load to the reinforced fibers [8]. The commonly used fibers are glass (in fiberglass), carbon, aramid, or basalt.

Fiber-reinforced composites are classified into continuous or discontinuous fiber-reinforced composites. The properties of fiber-reinforced composites mainly depend upon the type of fiber, length of the fiber, orientation, volume fraction, and the direction of an externally applied load [8]. For higher strength and stiffness, continuous-fiber composites are preferred. Fig. (4) shows the discontinuous and continuous fiber composites [7].

The fiber-reinforced composites may undergo structural failure due to the following reasons:

- The physical properties of composites are anisotropic. Therefore, loads at different locations may cause the failure of composites [7].
- Sometimes delamination occurs because of shock or stress.
- Compression failures can occur at a macro level or micro level in each

reinforcing fiber.
- Tension failures occur at a microscopic level when the layers in composites fail to form the bond between the matrix and fibers.

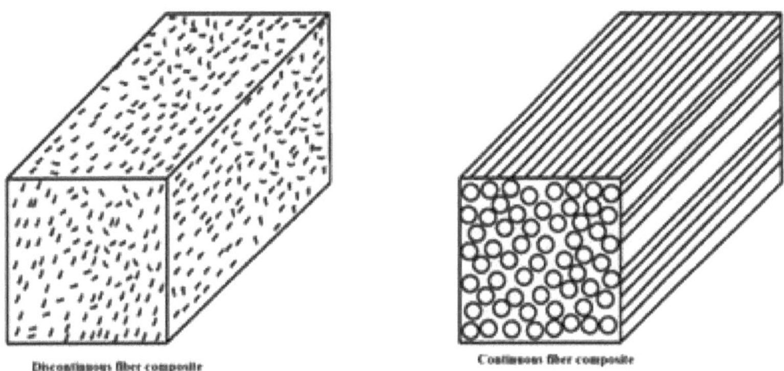


**Fig. (4).** Discontinuous and continuous fiber composites [7].

Properties of fiber-reinforced composites:

1. They are light in weight as compared to traditional building materials, such as concrete, metal, and wood.
2. They have high strength per unit of weight.
3. The reinforcing fiber prevents slip and crack propagation of the polymer matrix and improves the mechanical properties.
4. They are highly resistant to chemicals and they never rust nor corrode. They are also excellent heat resistants.
5. They are elastic and hence have higher yield strength, fracture strength, and fatigue life.
6. They offer greater design flexibility.
7. Some fiber-reinforced composites (*e.g.* fiberglass) are non-conductive and do not conduct electricity, and can be safely used in work involving electricity.

## 3. APPLICATIONS OF FIBER-REINFORCED COMPOSITES

1. Due to their lightweight, these composites are used in making cars and aircraft.
2. Due to their strength, composites like graphite-epoxy are used in making bridges.
3. Due to their resistance to corrosion, fiberglass is used in making cars and boats.
4. Since they are elastic, they are used in car leaf springs and the limbs of archery bows.
5. Because of their high strength, and insulation, these composites are used in

making armors.

6. Composites like fiberglass are used for thermal insulation and sound absorption.
7. Since they are non-conductive, ladders made with fiberglass are not risky even if they were to cross a power line.


## 3.1. Manufacturing Processes

There are many different manufacturing methods available to prepare different types of composites. The properties of the fabricated composites mainly depend upon the type of manufacturing process, experimental conditions, and the type of resin and reinforcements.


### 3.1.1. Spray Lay–up Process

In this process, the fibers are cut into small pieces using a chopper gun and fed into a catalyzed resin spray directed at the mold, as shown in Fig. (5) [9]. The deposited resins are allowed to cure under standard atmospheric conditions. Generally, polyester resin and glass roving reinforcements are used to prepare composites using this process. The prepared composites are mainly used for simple enclosures, caravan bodies, bathtubs, shower trays, and in truck fairings.

Some of the advantages of this method are:

1. It can be used for many years.
2. Easy and quick deposition of fibers and resin.
3. Low-cost tooling.

Disadvantages:

1. The prepared composites tend to be heavy due to the use of more resin.
2. Mechanical properties of the composite are limited due to the use of only short fibers.
3. Only low viscous resins can be used for the easy spray.
4. The use of low viscous resins can easily penetrate clothing and can cause skin diseases.

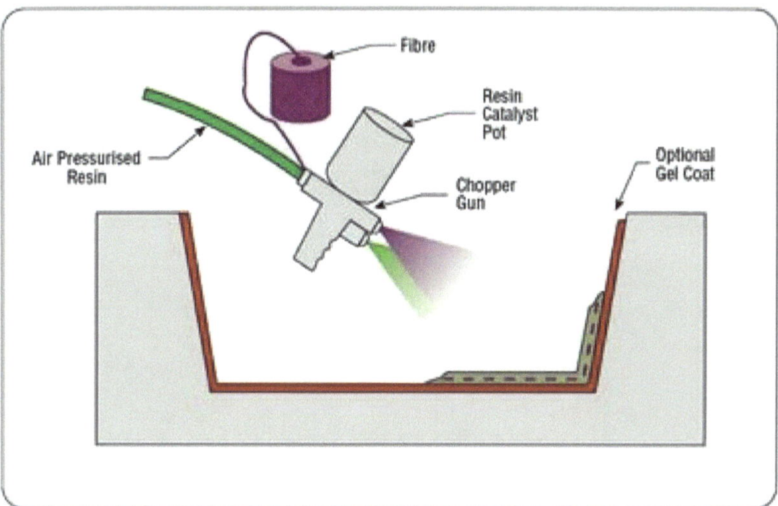


**Fig. (5).** Spray Lay–up process [9].

### 3.1.2. Wet Lay-up Process

This process is also called a hand lay-up process and is shown in Fig. (**6**) [9]. In this process, the resins are hand infused into woven or knitted fabric fibers. This is generally performed by rotating rollers on the resin bath containing reinforced fabrics. Laminates of composites are cured under standard atmospheric conditions. Epoxy, vinyl ester, phenolic, and polyester are used as resins and any type of fiber, including heavy aramid fabric fibers, can be used as reinforcement. The prepared composites are used in standard wind-turbine blades, production boats, and architectural moldings.

Advantages:

1. Long life
2. Involves simple principle
3. Inexpensive
4. Wide choice of material types.
5. Produces higher and longer fiber contents than the spray lay-up process.

Main Disadvantages:

1. This process requires highly skilled workers because the possibility of a huge amount of void formation is greater.
2. This process involves the use of low molecular weight resins and this is more

potential to be harmful.
3. The use of low viscous resin increases the tendency of penetrating clothes.
4. Low viscous resins affect the mechanical and thermal properties of the composites.

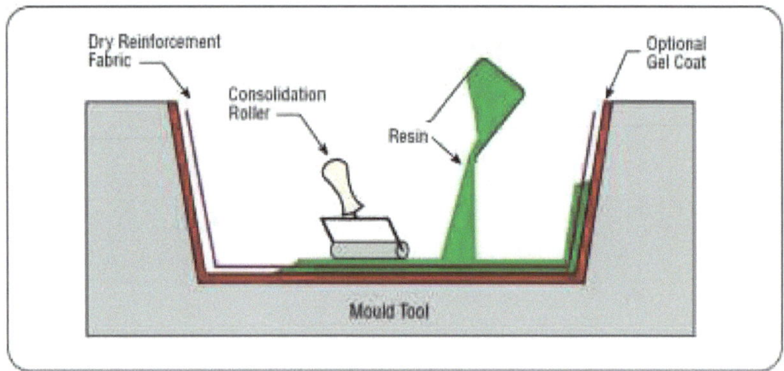


**Fig. (6).** Wet Lay-up process [9].

### 3.1.3. Vacuum Bagging

This process is just an improved version of the wet lay-up process, but here, laid-up is achieved by sealing a plastic film over the wet laid-up laminate and onto the tool. Fig. (7) shows the vacuum bagging process [9]. Epoxy and phenolic matrix composites can be fabricated easily and safely, but the use of polyesters and vinyl esters resins may have health issues due to the excessive extraction of styrene from the resin by the vacuum pump. Any type of fibers, including heavy fibers, can be reinforced by this method. The prepared composites are used in cruising boats, race car components, *etc.*

Advantages:

1. Preparation of high fiber content laminate composites can be achieved easily.
2. The prepared composites exhibit lesser voids than the composites prepared from the wet lay-up process.

Disadvantages:

1. Expensive.
2. A high skilled manpower is required

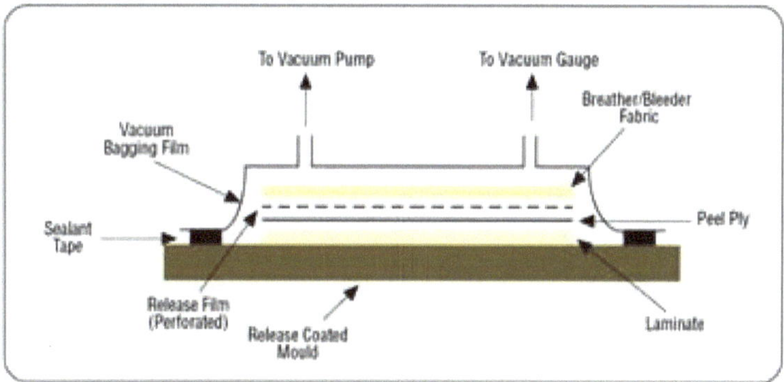



**Fig. (7).** Vacuum bagging process [9].

### *3.1.4. Filament Winding*

This process is generally used to prepare hollow, circular, or oval-shaped materials like pipes and tanks. Fig. (**8**) depicts the filament winding process [9]. Fiber tows are dipped in a resin bath and then wound onto a mandrel at different orientations.

Advantages:

1. This process is fast and economic
2. Resin content can be controlled.
3. No secondary operations are required.
4. Improved structural properties of laminates.

Disadvantages:

1. Limited to convex shaped materials.
2. The cost of Mandrel is more.

## 4. ADVANTAGES AND DISADVANTAGES OF COMPOSITES MATERIALS

### 4.1. Advantages [4]

- Excellent fatigue, impact, wear, and corrosion-resistant.
- High strength to weight ratio.
- No timely inspection is required.
- The fiber pattern can be utilized to increase mechanical properties.

- Improved dent resistance.
- Composites offer improved torsional stiffness.
- Thermoplastics can also be reformed.
- They are dimensionally stable.
- Easy and simple assembly and manufacturing process.

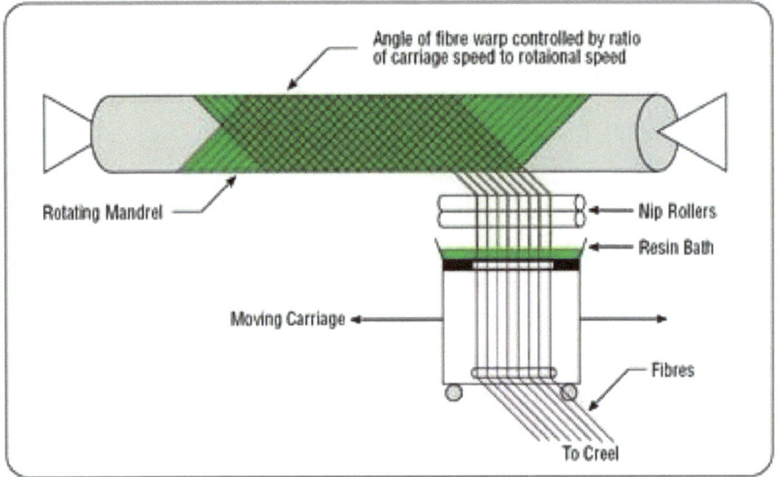


**Fig. (8).** Filament Winding process [9].

## 4.2. Disadvantages of Composites

- The cost of raw materials and fabrication is more.
- More prone to damage due to their brittle nature.
- Weak transverse properties.
- Matrix exhibit low toughness.
- Difficult to reuse.
- Hot curing is necessary and this is time-consuming.
- Analyses are difficult.
- The matrices undergo environmental degradation.

## 4.3. Nanocomposites

In recent years, nanocomposites have gained more popularity and became one of the hot research topics among researchers. Many research agencies and governments are sanctioning a lot of money to undertake research in nanocomposites. It is a separate class of composites having many advantages over traditional composite materials. Nanocomposites are the materials incorporated with nano-sized particles into a matrix of standard material. As a result of nano reinforcement, one can see the dramatic improvement as shown in a previous study [10];

- Mechanical properties
- Electrical conductivity
- Flame retardant properties
- Thermal stability
- Chemical resistance
- Surface appearance
- Optical properties
- Corrosion-resistant properties *etc.*

As we know, nanoparticles exhibit high surface energy and high surface area to volume ratio, which significantly changes their properties when compared with their bulk-sized equivalents [10]. The composition of the matrix is not going to change even though we add 0.5 to 5 weight% of nanoparticles. The addition of nanoparticles to the matrix can improve the properties of the composite at least by 1000 times compared to the addition of bulk-sized particles to the matrix. Materials like graphene, carbon nanotubes, fullerenes, nano metals, metal oxide nanoparticles, ceramic nanoparticles can be used as a reinforcement material to prepare nanocomposites.

## 5. WHY NANOCOMPOSITES EXHIBIT BETTER PROPERTIES THAN THEIR BULK COMPOSITES?

Nanosize of the reinforcement may result in the following:

a. Bulk composite materials exhibit weak physical sensitivity and result in poor thermal and mechanical properties. Whereas nanocomposites show a greater physical interaction between organic and inorganic materials and demonstrate excellent properties.

b. The higher chemical reactivity of nanocomposites at grain boundaries results in improved properties.

### 5.1. Classification of Nanocomposites

Nanocomposites are generally classified into the following:

1. Ceramic based nanocomposites

2. Metallic based nanocomposites
3. Polymer-based nanocomposites

### *5.1.1. Ceramic Based Nanocomposites*

In ceramic-based nanocomposites, ceramic materials act as a matrix and nanoparticles act as reinforcements. The combination of these 2 or more phases together constitutes ceramic nanocomposites. This increases the strength, hardness, and abrasion resistance of composite materials. Ceramic nanocomposites also exhibit enhanced ductility, toughness, formability, superplasticity, variation in electrical conduction, and magnetic properties.

### *5.1.2. Metallic Based Nanocomposites*

In metallic based nanocomposites, metallic materials act as a matrix and nanoparticles act as reinforcements. The combination of these 2 or more phases together constitutes metallic nanocomposites. Metallic nanocomposites show increased strength, hardness, electrical resistivity, magnetic properties, superplasticity, and decreased melting point.

### *5.1.3. Polymer-Based Nanocomposites*

In polymer-based nanocomposites, polymers act as a matrix, and nanoparticles act as reinforcements. The combination of these 2 or more phases together constitutes polymer nanocomposites. Nowadays, polymer nanocomposites are very famous among researchers due to the wide range of advantages over other nanocomposites.

## 6. RECENT ADVANCES IN COMPOSITES [11]

The recent past had witnessed ground-breaking innovations in the field of composites. Below are some of the new composite materials exhibiting different properties and applications:

a. Ceromers
b. Smart composites
c. Ormocers
d. Giomers

## 6.1. Ceromers

The term ceromer stands for Ceramic Optimized Polymer and was introduced by Ivoclar to describe their composite Tetric Ceram. They are micro filled hybrid resins consisting of a paste containing barium glass, spheroidal mixed oxide, ytterbium trifluoride, and silicon dioxide (57 vol%), dimethacrylate monomers [12]. Ceromers are mainly used for veneers without a metal framework, crowns, and bridges, including implant restorations on a metal framework.

## 6.2. Smart Composites

These are active dental polymers composed of bioactive amorphous calcium phosphate (ACP) filler capable of responding to environmental pH change by releasing calcium and phosphate ions [11]. Therefore, the name is smart composites and they are sometimes called intelligent composites.

## 6.3. Ormocer

Another new composite material called 'ormocer' was developed for dental restoration therapy by Dr. Herbert Wolters from the Fraunhofer Institute for Silicate Research. ORMOCER means 'Organically Modified Ceramic' which is a brand-new composite material used for all filling indications in the anterior and posterior area [13].

## 6.4. Giomers

These are newly developed hybrid aesthetic restorative materials for dental restorative therapy [11]. Giomers involve the use of pre-reacted glass ionomer (PRG) technology to form a stable composite material. Giomer has been called a light cure as it does not have a significant acid-base reaction as part of its curing process and cannot set in the dark.

Apart from the above, the world is witnessing a lot more improvement in the field of dental medicine like tooth-colored restoratives, and bonding technology have made dental procedures more palatable and feasible. Patients are attracted to a restoration that matches the color of natural teeth. These types of composites are high in demand and have become the most frequently used aesthetic material in dentistry.

## CONCLUSION

Composite materials are made up of two or more materials and their amalgamated combination can impart superior properties than those of individual components. Composite materials are preferred above some traditional materials for their lightweight, high strength, corrosion resistance, chemical resistance and oxidation resistance, and more affordable properties. These types of materials can be used in the construction of buildings, automobile body parts, aerospace industries, and defense industries, *etc.* Among all the composites, fiber-reinforced plastics are one of the important composite materials mainly used in aerospace, marine, automobile, construction, power industries, *etc.* Despite the high cost, fiber-reinforced plastics have become more popular in high-performance products for their excellent properties.

## QUESTIONS

1) What are composites?

2) Write the classification of composites based on matrix and reinforcement form.

3) What are organic matrix or polymer matrix composites? Write their advantages and disadvantages.

4) Write a note on a metal matrix and ceramic matrix composites.

5) What are laminar and particulate composites?

6) Explain the fiber-reinforced composites in detail.

7) How fiber-reinforced composites undergo structural failure?

8) Mention some of the important properties of fiber-reinforced composites.

9) Mention some of the applications of fiber-reinforced composites.

10) Explain different methods used to prepare composite materials.

11) What are the advantages and disadvantages of composites?

12) What are nanocomposites? How they are more advanced than normal composites?

13) Write a note on the classification of nanocomposites.

# REFERENCES

[1]     Wikipedia, "Composite material",  https://en.wikipedia.org/wiki/Composite_material

[2]     R.P.L. Nijssen, *Composite Materials-An introduction.* Inholland University of Applied Sciences, VKCN publications, 2015.

[3]     F.C. Campbell, Chapter 1, "Introduction to Composite Materials, Structural Composite Materials", *ASM International,* 2010.

[4]     NPTEL, "Introduction of Composites",  https://nptel.ac.in/courses/Webcourse-contents/IISc-BANG/ Composite%20Materials/pdf/Lecture_Notes/LNm1.pdf

[5]     A.H. Idrisi, and S. Deva, "Development and Testing of Metal Matrix Composite by Reinforcement of Sic Particles on Al 5XXX Series Alloy", *Int. J. Eng. Res. Technol. (Ahmedabad),* vol. 3, pp. 1303-1309, 2014. [IJERT].

[6]     Caterham Group, "Enlighten, Particulate reinforced composites",  https://altairenlighten.com/in-depth/particulate-reinforced-composites/

[7]     eFunda, "Introduction to Composite Materials",  https://www.efunda.com/formulae/solid_mechanics/ composites/comp_intro.cfm

[8]     S.V. Kailas, Chapter 12, "Composites, Material Science", *NPTEL lecture notes.* https://nptel.ac.in/courses/112108150/pdf/Lecture_Notes/MLN_12.pdf

[9]     "SP Systems, Guide to Composites, GTC-1-1098",  http://www.composites.ugent.be/home_ made_composites/documentation/SP_Composites_Guide.pdf

[10]    AZoNano, "Nanocomposites – An Overview of Properties, Applications, and Definition", https://www.azonano.com/article.aspx?ArticleID=1832

[11]    J. Marwaha, R. Goyal, Y. Sharma, and S. Mohanta, "Recent Advancement In Composites – A Review", *International Journal of Medical Science,* vol. 7, 2020.

[12]    S.C. Bayne, D.F. Taylor, and H.O. Heymann, "Protection hypothesis for composite wear", *Dent. Mater.,* vol. 8, no. 5, pp. 305-309, 1992.
        [http://dx.doi.org/10.1016/0109-5641(92)90105-L] [PMID: 1303372]

[13]    R.L. Bowen, F.C. Eichmiller, W.A. Marjenhoff, and N.W. Rupp, "Adhesive bonding of composites", *J. Am. Coll. Dent.,* vol. 56, no. 2, pp. 10-13, 1989.
        [PMID: 2528575]

# SUBJECT INDEX

## A

Abrasion resistance 178
Acid 8, 100, 101, 103
  adipic 100
  formic 103
  sebacic 101
Acts 54, 121, 123, 125, 141, 148, 149, 151, 153, 157, 158, 161, 169
  cement 169
  coatings 157
  metal oxide coating 141
  oxygen concentration 151
Actuators, engine manifold 132
Addition polymerization 88, 90, 91, 93, 95, 96, 97, 102, 109, 110
  anionic 93
  cationic 95
  free radical 96
  and condensation polymerization 109
  method 90
Advantages of polymer matrix composites 167
Aeration corrosion 151, 152
Aerobic bacteria 156
Alkali halides 44
Alkali metal 93
  alkyls 93
  amides 93
Alloys 1, 2, 51, 52, 53, 54, 61, 66, 67, 76, 113, 116, 133, 141, 142, 159, 162
  interstitial solid solution 66
  magnesium 76
  nickel-based 133
  preparing high strength metal 54
Aluminum alloys 76
Amorphous 7, 15, 16, 36, 179
  calcium phosphate (ACP) 179
  materials 7, 15, 16, 36
Anaerobic bacteria 156
Annealed copper alloys 76
Anodic 147, 148, 150, 151, 158, 159

coating 158
inhibitors 159
materials 150, 151, 158
reaction 147, 148, 150
Anti-corrosion coatings 158
Applications 1, 2, 3, 4, 5, 81, 96, 97, 99, 100, 101, 102, 103, 104, 105, 109, 110, 113, 131, 132, 134, 137, 138, 161, 167
  aeronautical 167
  aerospace 81, 132
  automobile 134
  automotive 131, 132
  of electroless plating 161
  of polyethylene 97
  of polypropylene 99
  of polystyrene 100
  of powder metallurgy 113, 131, 137, 138
Atomic packing 21, 23, 26, 29
  efficiency 21
  factor (APF) 21, 23, 26, 29
Atomization 116
  method 116
  process 116
Attack, electrochemical 150
Attraction 10, 11, 13
  direct Coulomb 10
  electrostatic 11
Audio cassettes 100
Automobile 141, 180
  body parts 180
  parts, old 141
Avogadro's number 32, 36

## B

Bacterial corrosion 156
Ball mills 113, 123, 124, 137
  low energy 124
  principle of planetary 123, 137
Blending technology 125
Body-centered cubic structure 24, 25
Boltzmann's constant 41

Weak transverse properties 176

# Y

Young's modulus 65

# Z

Ziegler-Natta catalyst 87, 97, 98
Zinc coating 151, 158